NONSTANDARD FINITE DIFFERENCE MODELS OF DIFFERENTIAL EQUATIONS

NONSTANDARD FINITE DIFFERENCE MODELS OF DIFFERENTIAL EQUATIONS

Ronald E. Mickens

Callaway Professor of Physics
Clark Atlanta University

World Scientific
Singapore • New Jersey • London • Hong Kong

Published by

World Scientific Publishing Co. Pte. Ltd.

5 Toh Tuck Link, Singapore 596224

USA office: 27 Warren Street, Suite 401-402, Hackensack, NJ 07601

UK office: 57 Shelton Street, Covent Garden, London WC2H 9HE

Library of Congress Cataloging-in-Publication Data
Mickens, Ronald E., 1943–
   Nonstandard finite difference models of differential equations /
Ronald E. Mickens.
     p. cm.
   Includes bibliographical references and index.
   ISBN-13 9789810214586 -- ISBN-10 9810214588
   1. Finite differences. 2. Differential equations -- Numerical
solutions. I. Title.
QA431.M428 1994
515'.624--dc20   93-37665
                          CIP

British Library Cataloguing-in-Publication Data
A catalogue record for this book is available from the British Library.

This book is dedicated
to my wife
Maria,
my son
James Williamson,
my daughter
Leah maria.

Preface

This book was written in response to a large number of requests for copies of the author's papers on nonstandard finite difference schemes for the numerical integration of differential equations. The book provides a general summary of the methods used for the construction of such schemes. The major goal is to show that discrete (finite-difference) models exist for which the elementary types of numerical instabilities do not occur. The guiding philosophy behind this work is to get the qualitative details correct while not being overly concerned, at this level of the analysis, with the quantitative numerical results. (In any case, for most applications, the values of the various step-sizes are generally determined by the physical scales of the particular phenomena being studied.) The theoretical basis of our nonstandard discrete modeling methods is centered at the concepts of "exact" and "best" finite difference schemes. A set of rules is presented for constructing nonstandard finite difference schemes. The application of these rules often leads to an "essentially" unique finite difference model for a particular differential equation. It is expected that additional rules and restrictions will be discovered as research proceeds in this area.

An important feature of this book is the illustration of the various discrete modeling principles by their application to a large number of both ordinary and partial differential equations. The background requirements needed to fully understand the text are satisfied by the knowledge acquired in an introductory course on the numerical integration of differential equations.

I thank my many colleagues for their interest in my work. Again, I am particularly grateful to Ms. Annette Rohrs for typing the complete manuscript. Both she and Maria Mickens provided valuable editorial assistance. Finally, I thank the National Aeronautics and Space Administration for providing funds that allowed me to do research on nonstandard finite difference schemes.

Ronald E. Mickens
Atlanta, Georgia
August 1993

Table of Contents

Chapter 1

INTRODUCTION

1.1 Numerical Integration

In general, a given linear or nonlinear differential equation does not have a complete solution that can be expressed in terms of a finite number of elementary functions [1–4]. A first attack on this situation is to seek approximate analytic solutions by means of various perturbation methods [5–7]. However, such procedures only hold for limited ranges of the (dimensionless) system parameters and/or the independent variables. For arbitrary values of the system parameters, at the present time, only numerical integration techniques can provide accurate numerical solutions to the original differential equations of interest. A major difficulty with numerical techniques is that a separate calculation must be formulated for each particular set of initial and/or boundary values. Consequently, obtaining a global picture of the general solution to the differential equations often requires a great deal of computation and time. However, for many problems currently being investigated in science and technology, there exist no alternatives to numerical integration.

The process of numerical integration is the replacement of a set of differential equations, both of whose independent and dependent variables are continuous, by a model for which these variables may be discrete. In general, in the model the independent variables have a one-to-one correspondence with the integers, while the dependent variables can take real values. Our major concern in this book will be the use of a particular technique for constructing discrete models of differential equations, namely, the use of finite-difference methods [8–13]. No other procedures will be considered.

An important fact often overlooked in the formulation of discrete models of differential equations is that numerical integration methods should always be constructed with the help of the knowledge gained from the study of special solutions of

2

the differential equations. For example, if the differential equations have a constant solution with a particular stability property, the discrete model should also have this constant solution with exactly the same stability property [12, 13, 14]. We will consider this issue in considerable detail in Chapters 2 and 3.

1.2 Standard Finite-Difference Modeling Rules

To illustrate the construction of discrete finite-difference models of differential equations, we begin with the scalar ordinary equation

$$\frac{dy}{dt} = f(y), \tag{1.2.1}$$

where $f(y)$ is, in general, a nonlinear function of y. For a uniform lattice, with step-size, $\Delta t = h$, we replace the independent variable t by

$$t \to t_k = hk, \tag{1.2.2}$$

where k is an integer, i.e.,

$$t \in \{\ldots, -2, -1, 0, 1, 2, 3, \ldots\}. \tag{1.2.3}$$

The dependent variable $y(t)$ is replaced by

$$y(t) \to y_k, \tag{1.2.4}$$

where y_k is the approximation of $y(t_k)$. Likewise, the function $f(y)$ is replaced by

$$f(y) \to f_k, \tag{1.2.5}$$

where f_k is the approximation to $f[y(t_k)]$. The simplest possibility for f_k is

$$f_k = f(y_k). \tag{1.2.6}$$

For the first derivative, any one of the following forms is suitable

$$\frac{dy}{dt} \rightarrow \begin{cases} \frac{y_{k+1}-y_k}{h}, \\ \frac{y_k-y_{k-1}}{h}, \\ \frac{y_{k+1}-y_{k-1}}{2h}. \end{cases} \qquad (1.2.7)$$

These representations of the discrete first derivative are known, respectively, as the forward Euler, backward Euler, and central difference schemes. They follow directly from the conventional definition of the first derivative as given in a standard first course in calculus [15], i.e.,

$$\frac{dy}{dt} = \lim_{h \to 0} \begin{cases} \frac{y(t+h)-y(t)}{h}, \\ \frac{y(t)-y(t-h)}{h}, \\ \frac{y(t+h)-y(t-h)}{2h}. \end{cases} \qquad (1.2.8)$$

Given a first order scalar ordinary differential equation, a discrete finite-difference model is constructed by replacing in Eq. (1.2.1) the corresponding discrete terms of Eqs. (1.2.2) to (1.2.7). Thus, a simple finite-difference model for Eq. (1.2.1) is given by the expression

$$\frac{y_{k+1}-y_k}{h} = f(y_k). \qquad (1.2.9)$$

Other, at this stage of our discussion, equally valid discrete models are

$$\frac{y_k - y_{k-1}}{h} = f(y_k), \qquad (1.2.10)$$

$$\frac{y_{k+1}-y_{k-1}}{2h} = f(y_k), \qquad (1.2.11)$$

$$\frac{y_{k+1}-y_{k-1}}{2h} = f\left(\frac{y_{k+1}+y_{k-1}}{2}\right). \qquad (1.2.12)$$

The model of Eq. (1.2.10) is the backward Euler scheme. It is called an implicit scheme since for general nonlinear $f(y)$, y_k must be solved for at each value of k in terms of the previous y_{k-1}. The Eq. (1.2.11) gives the corresponding central difference scheme, while Eq. (1.2.12) is a mixed implicit, central difference scheme.

4

Note that all of the discrete representations reduce to the original differential equation in the appropriate limit [8].

$$h \to 0, \qquad k \to \infty, \qquad t_k = t = \text{fixed.} \tag{1.2.13}$$

These results indicate that the discrete modeling process has a great deal of non-uniqueness built into it.

For completeness, we give the standard discrete representation for the second derivative; it is [8]

$$\frac{d^2y}{dt^2} \to \frac{y_{k+1} - 2y_k + y_{k-1}}{h^2}. \tag{1.2.14}$$

Again, it follows directly from the standard calculus definition of the second derivative [15].

In the next section, we will use these standard finite-difference modeling rules to construct discrete representations for several rather simple, but, important in many applications, ordinary and partial differential equations.

1.3 Examples

All of the differential equations to be modeled in this section have been put in dimensional form. This means that all non-essential constants and parameters that arise in the original differential equations have been eliminated. We show how this can be done by considering two of these equations: the decay and Logistic equations. The general procedure is detailed in Mickens [6].

The decay equation is

$$\frac{dx}{dt} = -\lambda x, \qquad x(0) = x_0 = \text{given,} \tag{1.3.1}$$

where λ is a positive constant. Let $\bar{t}$ and y be defined as

$$\bar{t} = \lambda t, \qquad y = \frac{x}{x_0}. \tag{1.3.2}$$

Substitution of these results into Eq. (1.3.1) gives the dimensionless equation

$$\frac{dy}{d\bar{t}} = -y, \qquad y(0) = 1. \tag{1.3.3}$$

The so-called Logistic differential equation is

$$\frac{dx}{dt} = \lambda_1 x - \lambda_2 x^2, \qquad x(0) = x_0 = \text{given}, \tag{1.3.4}$$

where λ_1 and λ_2 are positive constants. This equation can be rewritten to the form

$$\frac{dx}{\lambda_1 dt} = x \left[1 - \left(\frac{\lambda_2}{\lambda_1} \right) x \right]. \tag{1.3.5}$$

Now let

$$\bar{t} = \lambda_1 t, \qquad y = \left(\frac{\lambda_2}{\lambda_1} \right) x. \tag{1.3.6}$$

Substitution of Eq. (1.3.6) into Eq. (1.3.5) gives the following dimensionless equation

$$\frac{dy}{d\bar{t}} = y(1 - y), \qquad y(0) = \left(\frac{\lambda_2}{\lambda_1} \right) x_0. \tag{1.3.7}$$

Observe that in dimensionless form, both the decay and Logistic equations have no arbitrary parameters.

Independently, as to whether we wish to numerically integrate a differential equation or not, it should always be transformed to a dimensionless form. Note that the "physical" original differential equation connects the derivatives of a physical variable such as distance or current and its relations to the various physical parameters, while the dimensionless transformed equation relates the various derivatives of a "mathematical" variable and associated constants that appear in the equation.

1.3.1 Decay Equation

The decay differential equation is

$$\frac{dy}{dt} = -y. \tag{1.3.8}$$

The direct forward Euler scheme is

$$\frac{y_{k+1} - y_k}{h} = -y_k.$$ (1.3.9)

A discrete model can also be constructed by using a symmetric expression for the linear term y in Eq. (1.3.8). For example

$$\frac{y_{k+1} - y_k}{h} = -\left(\frac{y_{k+1} + y_k}{2}\right).$$ (1.3.10)

The use of the central difference for the first derivative gives

$$\frac{y_{k+1} - y_{k-1}}{2h} = -y_k.$$ (1.3.11)

Another central difference scheme is

$$\frac{y_{k+1} - y_{k-1}}{2h} = -\left(\frac{y_{k+1} + y_k + y_{k-1}}{3}\right).$$ (1.3.12)

Likewise, a backward Euler model is

$$\frac{y_k - y_{k-1}}{h} = -y_k,$$ (1.3.13)

which can be written as

$$\frac{y_{k+1} - y_k}{h} = -y_{k+1}.$$ (1.3.14)

It is clear that these discrete models are all different. For example, writing them in reduced form gives the following results for the indicated equations: Eq. (1.3.9):

$$y_{k+1} = (1 - h)y_k,$$ (1.3.15)

Eq. (1.3.10):

$$y_{k+1} = \frac{(1 - h/2)}{(1 + h/2)} y_k,$$ (1.3.16)

Eq. (1.3.11):

$$y_{k+2} + 2hy_{k+1} - y_k = 0,$$ (1.3.17)

Eq. (1.3.12):

$$\left(1 + \frac{2h}{3}\right) y_{k+2} + \left(\frac{2h}{3}\right) y_{k+1} - \left(1 - \frac{2h}{3}\right) y_k = 0, \qquad (1.3.18)$$

Eq. (1.3.14):

$$y_{k+1} = \left(\frac{1}{1+h}\right) y_k. \qquad (1.3.19)$$

Note that Eqs. (1.3.15), (1.3.16) and (1.3.19) are first-order linear difference equations, while Eqs. (1.3.17) and (1.3.18) are second-order linear difference equations. Further, observe that all the equations of a given order have different constant coefficients for fixed step-size h. This implies that Eqs. (1.3.16) to (1.3.19) have different solutions [16]. Consequently, we must conclude that each of the above discrete models of the decay equation gives unique numerical solutions that differ from that of the other discrete models.

Again observe that each of these discrete models has coefficients that depend on the step-size h. This leaves open the possibility that the solution behaviors may vary with h. In the next chapter, we will see that this is in fact the situation.

1.3.2 Logistic Equation

The forward Euler scheme for the Logistic differential equation

$$\frac{dy}{dt} = y(1-y), \qquad (1.3.20)$$

is

$$\frac{y_{k+1} - y_k}{h} = y_k(1 - y_k). \qquad (1.3.21)$$

The corresponding backward Euler and central difference schemes are, respectively,

$$\frac{y_{k+1} - y_k}{h} = y_{k+1}(1 - y_{k+1}), \qquad (1.3.22)$$

$$\frac{y_{k+1} - y_{k-1}}{2h} = y_k(1 - y_k). \qquad (1.3.23)$$

The above three equations can be rewritten to the following forms:

Eq. (1.3.21):

$$y_{k+1} = (1+h)y_k - h(y_k)^2, \tag{1.3.24}$$

Eq. (1.3.22):

$$h(y_{k+1})^2 + (1-h)y_{k+1} - y_k = 0, \tag{1.3.25}$$

Eq. (1.3.23):

$$y_{k+2} = y_k + 2hy_{k+1}(1 - y_{k+1}). \tag{1.3.26}$$

Examination of these three equations shows that while all of them are nonlinear difference equations, the forward and backward Euler schemes are first-order, while the central scheme is second-order. The forward Euler and the central schemes are explicit, in the sense that the value of y_k can be determined from its, respective, values at y_{k-1} and, y_{k-1} and y_{k-2}. However, the backward Euler scheme requires the solution of a quadratic equation at each step. The existence and uniqueness theorems for difference equations [16] lead to the conclusion that these three finite-difference schemes for the Logistic equation have different solutions and the nature of the solutions may change as a function of the step-size h.

1.3.3 Harmonic Oscillator

The dimensionless, damped harmonic oscillator equation is

$$\frac{d^2y}{dt^2} + 2\epsilon\frac{dy}{dt} + y = 0. \tag{1.3.27}$$

Consider first the case for which $\epsilon = 0$, i.e., no damping is present. For this situation the equation of motion is

$$\frac{d^2y}{dt^2} + y = 0. \tag{1.3.28}$$

The simplest discrete model is one that uses a central difference for the discrete second derivative; this scheme is

$$\frac{y_{k+1} - 2y_k + y_{k-1}}{h^2} + y_k = 0. \tag{1.3.29}$$

Two other models that use a symmetric form for the linear term y in the differential equation are

$$\frac{y_{k+1} - 2y_k + y_{k-1}}{h^2} + \frac{y_{k+1} + y_{k-1}}{2} = 0, \tag{1.3.30}$$

and

$$\frac{y_{k+1} - 2y_k + y_{k-1}}{h^2} + \frac{y_{k+1} + y_k + y_{k-1}}{3} = 0. \tag{1.3.31}$$

Two discrete models having nonsymmetric forms for the linear term y are the following

$$\frac{y_{k+1} - 2y_k + y_{k-1}}{h^2} + y_{k-1} = 0, \tag{1.3.32}$$

$$\frac{y_{k+1} - 2y_k + y_{k-1}}{h^2} + y_{k+1} = 0. \tag{1.3.33}$$

All four models are linear, second-order difference equations with constant (for $h =$ fixed) coefficients. These coefficients differ from one model to the next. Consequently, we again must conclude that each discrete model will provide a different numerical solution.

The same conclusion is reached when the damping term in Eq. (1.3.27) is present, i.e., $\epsilon > 0$. For example, the following three equations correspond to using a centered discrete second-order derivative, a centered linear term, and, respectively, forward Euler, backward Euler and centered representations for the discrete first-order derivative:

$$\frac{y_{k+1} - 2y_k + y_{k-1}}{h^2} + 2\epsilon\left(\frac{y_{k+1} - y_k}{h}\right) + y_k = 0, \tag{1.3.34}$$

$$\frac{y_{k+1} - 2y_k + y_{k-1}}{h^2} + 2\epsilon\left(\frac{y_k - y_{k-1}}{h}\right) + y_k = 0, \tag{1.3.35}$$

$$\frac{y_{k+1} - 2y_k + y_{k-1}}{h^2} + 2\epsilon\left(\frac{y_{k+1} - y_{k-1}}{2h}\right) + y_k = 0. \tag{1.3.36}$$

All of these models are second-order and linear; however, they clearly have different constant coefficients which implies that they have different solutions.

1.3.4 Unidirectional Wave Equation

A linear equation that describes waves propagating along the x-axis with unit velocity is the unidirectional wave equation

$$u_t + u_x = 0, \tag{1.3.37}$$

where $u = u(x,t)$ and

$$u_t \equiv \frac{\partial u}{\partial t}, \qquad u_x \equiv \frac{\partial u}{\partial x}. \tag{1.3.38}$$

Denote the discrete space and time variables by

$$t_k = (\Delta t)k, \qquad x_m = (\Delta x)m, \tag{1.3.39}$$

where

$$k \in \{\ldots, -2, -1, 0, 1, 2, 3, \ldots\}, \tag{1.3.40}$$

$$m \in \{\ldots, -2, -1, 0, 1, 2, 3, \ldots\}. \tag{1.3.41}$$

Thus, the discrete approximation to $u(x,t)$ is

$$u(x,t) \to u_m^k, \tag{1.3.42}$$

and the corresponding discrete first-derivatives are [8, 12]

$$\frac{\partial u}{\partial t} \to \begin{cases} \frac{u_m^{k+1} - u_m^k}{\Delta t}, \\ \frac{u_m^k - u_m^{k-1}}{\Delta t} \\ \frac{u_m^{k+1} - u_m^{k-1}}{2\Delta t}, \end{cases} \tag{1.3.43}$$

and

$$\frac{\partial u}{\partial x} \to \begin{cases} \frac{u_{m+1}^k - u_m^k}{\Delta x}, \\ \frac{u_m^k - u_{m-1}^k}{\Delta x} \\ \frac{u_{m+1}^k - u_{m-1}^k}{2\Delta x}. \end{cases} \tag{1.3.44}$$

Various discrete models can be constructed by selecting a particular representation for the discrete time-derivative and a second particular representation for the discrete space-derivative. The following four cases illustrate this procedure.

(i) Forward Euler time-derivative and forward Euler space-derivative:

$$\frac{u_m^{k+1} - u_m^k}{\Delta x} + \frac{u_{m+1}^k - u_m^k}{\Delta x} = 0;$$ (1.3.45)

(ii) forward Euler time-derivative and backward Euler space-derivative:

$$\frac{u_m^{k+1} - u_m^k}{\Delta t} + \frac{u_m^k - u_{m-1}^k}{\Delta x} = 0;$$ (1.3.46)

(iii) forward Euler time-derivative and central difference space-derivative:

$$\frac{u_m^{k+1} - u_m^k}{\Delta t} + \frac{u_{m+1}^k - u_{m-1}^k}{2\Delta x} = 0;$$ (1.3.47)

(iv) an implicit scheme with forward Euler for the time-derivative and backward Euler for the space-derivative:

$$\frac{u_m^{k+1} - u_m^k}{\Delta t} + \frac{u_m^{k+1} - u_{m-1}^{k+1}}{\Delta x} = 0.$$ (1.3.48)

Clearly, a number of other discrete models can be easily constructed.

All of the above equations are linear partial difference equations with constant coefficients (for fixed Δt and Δx). The models of Eqs. (1.3.45), (1.3.46) and (1.3.48) are of first-order in both the discrete time and space variables. Equation (1.3.47) is first-order in the discrete time variable, but, is of second-order in the discrete space variable. Again, by inspection all four models are different and thus will give different numerical solutions to the original partial differential equation.

1.3.5 Diffusion Equation

The simple linear diffusion partial differential equation, in dimensionless form, is

$$u_t = u_{xx}, \qquad u = u(x,t).$$ (1.3.49)

The standard explicit form for this equation is given by the expression

$$\frac{u_m^{k+1} - u_m^k}{\Delta t} = \frac{u_{m+1}^k - 2u_m^k + u_{m-1}^k}{(\Delta x)^2},$$ (1.3.50)

while the standard implicit form is

$$\frac{u_m^{k+1} - u_m^k}{\Delta t} = \frac{u_{m+1}^{k+1} - 2u_m^{k+1} + u_{m-1}^{k+1}}{(\Delta x)^2}. \tag{1.3.51}$$

While both of these equations are linear, partial difference equations that are first-order in the discrete time and second-order in the discrete space variables, they are not identical and consequently their solutions will give different numerical solutions to the diffusion equation.

1.3.6 Burgers' Equation

The inviscid Burgers' partial differential equation is [13]

$$u_t + u u_x = 0, \qquad u = u(x, t). \tag{1.3.52}$$

The following four equations are examples of discrete models that can be constructed for Eq. (1.3.52) using the standard finite-difference rules.

(i) Forward Euler for the time-derivative and forward Euler for the space-derivative:

$$\frac{u_m^{k+1} - u_m^k}{\Delta t} + u_m^k \left(\frac{u_{m+1}^k - u_m^k}{\Delta x} \right) = 0; \tag{1.3.53}$$

(ii) forward Euler for the time-derivative and implicit forward Euler for the space-derivative:

$$\frac{u_m^{k+1} - u_m^k}{\Delta t} + u_m^k \frac{u_{m+1}^{k+1} - u_m^{k+1}}{\Delta x} = 0; \tag{1.3.54}$$

(iii) central difference schemes for both the time- and space-derivatives:

$$\frac{u_m^{k+1} - u_m^{k-1}}{2\Delta t} + u_m^k \left(\frac{u_{m+1}^k - u_{m-1}^k}{2\Delta x} \right) = 0; \tag{1.3.55}$$

(iv) forward Euler for the time-derivative and backward Euler for the space-derivative:

$$\frac{u_m^{k+1} - u_m^k}{\Delta x} + u_m^k \left(\frac{u_m^k - u_{m-1}^k}{\Delta x} \right) = 0. \tag{1.3.56}$$

Note that in the limits

$$k \to \infty, \qquad \Delta t \to 0, \qquad t_k = t = \text{fixed}, \qquad (1.3.57)$$

$$m \to \infty, \qquad \Delta x \to 0, \qquad x_k = x = \text{fixed}, \qquad (1.3.58)$$

all of these difference schemes reduce to the inviscid Burgers' equation. However, inspection shows that for finite Δt and Δx these four partial difference equations are not identical. This fact leads to the conclusion that they will give numerical solutions that differ from each other.

1.4 Critique

The major result coming from the analysis of the previous two sections is the ambiguity of the modeling process for the construction of discrete finite-difference models of differential equations. The use of the standard rules does not lead to a unique discrete model. Consequently, one of the questions before us is which, if any, of the standard finite-difference schemes should be used to obtain numerical solutions for a particular differential equation? Another very important issue is the relationship between the solutions to a given discrete model and that of the corresponding differential equation. As indicated in Section 1.3, this connection may be tenuous. This and related matters lead to the study of numerical instabilities which is the subject of Chapter 2.

Once a discrete model is selected, the calculation of a numerical solution requires the choice of a time and/or space step-size. How should this be done? For problems in the sciences and engineering, the value of the step-sizes must be determined such that the physical phenomena of interest can be resolved on the scale of the computational grid or lattice [12–14]. However, suppose one is interested in the long-time or asymptotic-in-space behavior of the solution; can the step-sizes be taken as large as one wishes?

The general issue can be summarized as follows: Consider the scalar differential equation of Eq. (1.2.1). Select a finite-difference scheme to numerically integrate this equation. At the grid point $t = t_k$, denote by $y(t_k)$ the solution to the differential equation and by $y_k(h)$ the solution to the discrete model. (Note that the numerical solution is written in such a way as to indicate that its value depends on the step-size, h.) What is the relationship between $y(t_k)$ and $y_k(h)$? In particular, how does this relationship change as h varies?

Numerical instabilities exist in the numerical solutions whenever the qualitative properties of $y_k(h)$ differ from those of $y(t_k)$. One of the tasks of this book will be to eliminate the elementary numerical instabilities that can arise in the finite-difference models of differential equations. Our general goal will be the construction of discrete models whose solutions have the same qualitative properties as that of the corresponding differential equation for all step-sizes. We have not entirely succeeded in this effort, but, progress has definitely been made.

A final comment should be made on the issue of chaos and differential equations. In the past several decades, much effort has been devoted to the study of "mathematical chaos" as it occurs in the solutions of deterministic systems modeled by coupled differential equations [17–19]. Experimentally, chaotic-like behavior has been measured in fluid phenomena [19, 20], chemical reactions [21], nonlinear electrical and mechanical oscillations [22, 23] and in biomedical systems [24]. In this book, we do not address these issues. Our task is to formulate discrete finite-difference models of differential equations that have numerical solutions which reflect accurately the underlying mathematical structures of the solutions to the differential equations.

References

1. S. L. Ross, *Differential Equations* (Blaisdell; Waltham, MA; 1964).

2. M. Humi and W. Miller, *Second Course in Ordinary Differential Equations for Scientists and Engineers* (Springer-Verlag, New York, 1988).

3. D. Zwillinger, *Handbook of Differential Equations* (Academic Press, Boston, 1989).

4. D. Zwillinger, *Handbook of Integration* (Jones and Bartlett, Boston, 1992).

5. C. M. Bender and S. A. Orszag, *Advanced Mathematical Methods for Scientists and Engineers* (McGraw-Hill, New York, 1978).

6. R. E. Mickens, *Nonlinear Oscillations* (Cambridge University Press, New York, 1981).

7. J. Kevorkian and J. D. Cole, *Perturbation Methods in Applied Mathematics* (Springer-Verlag, New York, 1981).

8. F. B. Hildebrand, *Finite-Difference Equations and Simulations* (Prentice-Hall; Englewood Cliffs, NJ; 1968).

9. M. K. Jain, *Numerical Solution of Differential Equations* (Halsted Press/John Wiley and Sons, New York, 2nd edition, 1984).

10. J. M. Ortega and W. G. Poole, Jr., *An Introduction to Numerical Methods for Differential Equations* (Pitman; Marshfield, MA; 1981).

11. J. C. Butcher, *The Numerical Analysis of Ordinary Differential Equations: Runge-Kutta and General Linear Methods* (Wiley-Interscience, New York, 1987).

12. D. Greenspan and V. Casulli, *Numerical Analysis for Applied Mathematics, Science and Engineering* (Addison-Wesley; Redwood City, CA; 1988).

13. M. B. Allen III, I. Herrera and G. F. Pinder, *Numerical Modeling in Science and Engineering* (Wiley-Interscience, New York, 1988).

14. D. Potter, *Computational Physics* (Wiley-Interscience, New York, 1973).

15. J. Marsden and A. Weinstein, *Calculus I* (Springer-Verlag, New York, 1984); section 1.3.

16. R. E. Mickens, *Difference Equations: Theory and Applications* (Van Nostrand Reinhold, New York, 2nd edition, 1990).

17. S. Wiggins, *Introduction to Applied Nonlinear Dynamical Systems and Chaos* (Springer-Verlag, New York, 1990).

18. R. L. Devaney, *An Introduction to Chaotic Dynamical Systems* (Benjamin/ Cummings; Menlo Park, CA; 1986).

19. Hao Bai-Lin, editor, *Chaos* (World Scientific, Singapore, 1984).

20. G. L. Baker and J. P. Gollub, *Chaotic Dynamics* (Cambridge University Press, New York, 1990).

21. S. K. Scott, *Chemical Chaos* (Clarendon Press, Oxford, 1991).

22. F. C. Moon, *Chaotic Vibrations* (Wiley-Interscience, New York, 1987).

23. J. M. T. Thompson and H. B. Stewart, *Nonlinear Dynamics and Chaos* (Wiley, New York, 1986).

24. B. J. West, *Fractal Physiology and Chaos in Medicine* (World Scientific, Singapore, 1990).

Chapter 2
NUMERICAL INSTABILITIES

2.1 Introduction

A discrete model of a differential equation is said to have numerical instabilities if there exist solutions to the finite-difference equations that do not correspond qualitatively to any of the possible solutions of the differential equation. It is doubtful if a precise definition can ever be stated for the general concept of numerical instabilities. This is because it is always possible, in principle, for new forms of numerical instabilities to arise when new nonlinear differential equations are discretely modeled. The concept, as we will use it in this book, will be made clearer in the material to be discussed in this Chapter. Numerical instabilities are an indication that the discrete equations are not able to model the correct mathematical properties of the solutions to the differential equations of interest.

The fundamental reason for the existence of numerical instabilities is that the discrete models of differential equations have a larger parameter space than the corresponding differential equations. This can be easily demonstrated by the following argument. Assume that a given dynamic system is described in terms of the differential equation

$$\frac{dy}{dt} = f(y, \lambda) \tag{2.1.1}$$

where λ denotes the n-dimensional parameter vector that defines the system. A discrete model for Eq. (2.1.1) takes the form

$$y_{k+1} = F(y_k, \lambda, h) \tag{2.1.2}$$

where $h = \Delta t$ is the time step-size. Note that the function F contains $(n+1)$ parameters; this is because h can now be regarded as an additional parameter. The solutions to Eqs. (2.1.1) and (2.1.2) can be written, respectively as $y(t, \lambda)$ and $y_k(\lambda, h)$. Even if $y(t, \lambda)$ and $y_k(\lambda, h)$ are "close" to each other for a particular value of h, say $h = h_1$. If h is changed to a new value, say $h = h_2$, the possibility exists that $y_k(\lambda, h_2)$ differs greatly from $y_k(\lambda, h_1)$ both qualitatively and quantitatively. The detailed study of what actually occurs relies on the use of bifurcation theory [1, 2, 3].

The purpose of this Chapter is to consider a number of differential equations, construct several discrete models of them, and compare the properties of the solutions to the difference equations to the corresponding properties of the original differential equations. Any discrepancies found are indications of numerical instabilities. We end the Chapter with a summary and discussion of the elementary numerical instabilities.

2.2 Decay Equation

The decay differential equation is

$$\frac{dy}{dt} = -y. \qquad (2.2.1)$$

Even if we did not know how to solve this equation exactly, its general solution behavior could be obtained by knowledge of the fact that for $y > 0$, the derivative is negative, while for $y < 0$, the derivative is positive. Also, $y = 0$ is a solution of the differential equation. Consequently, all solutions have the forms as shown in Figure 2.2.1(b). For the initial condition

$$y(t_0) = y_0, \qquad (2.2.2)$$

the exact general solution is

$$y(t) = y_0 e^{-(t-t_0)}. \qquad (2.2.3)$$

From either Figure 2.2.1(b) or Eq. (2.2.3), it can be concluded that all solutions monotonically decrease (in absolute value) to zero as $t \to \infty$.

The forward Euler scheme for the decay equation is

$$\frac{y_{k+1} - y_k}{h} = -y_k, \qquad (2.2.4)$$

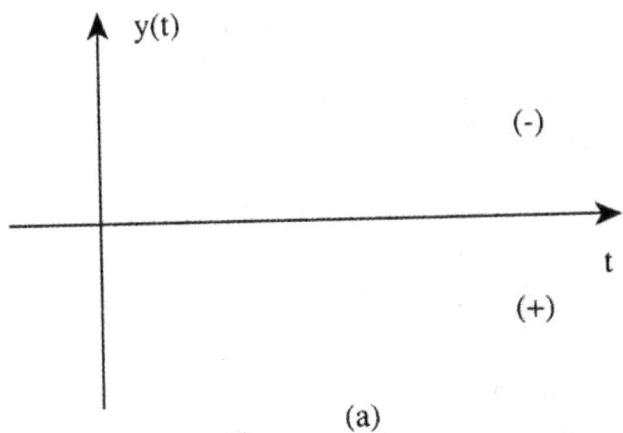

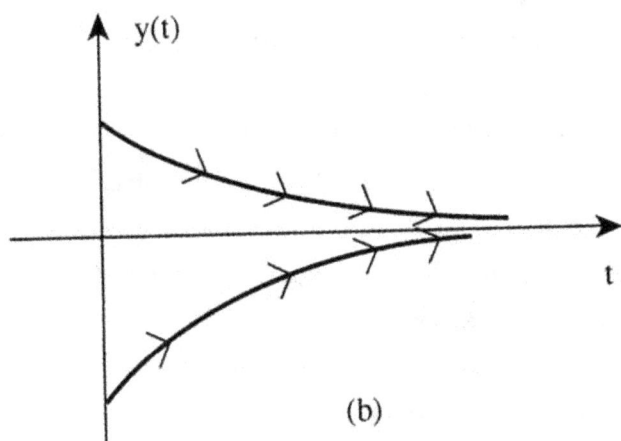

Figure 2.2.1. The decay equation. (a) Regions where the derivative has a constant sign. (b) Typical trajectories.

where $h = \Delta t$ is the step-size. It can be rewritten to the form

$$y_{k+1} = (1 - h)y_k, \tag{2.2.5}$$

which shows it to be a first-order, linear difference equation with constant coefficients. Its solution is

$$y_k = y_0(1 - h)^k. \tag{2.2.6}$$

Note that the behavior of the solution depends on the value of $r(h) = 1 - h$ which is plotted in Figure 2.2.2. Referring to Figure 2.2.3, the following conclusions are reached:

 (i) If $0 < h < 1$, then y_k decreases monotonically to zero.

 (ii) If $h = 1$, then for $k \geq 1$, the solution is identically zero.

 (iii) If $1 < h < 2$, y_k decreases to zero with an oscillating (change in sign) amplitude.

 (iv) If $h = 2$, then y_k oscillates with a constant amplitude. The solution has period-two.

 (v) If $h > 2$, y_k oscillates with an exponentially increasing amplitude.

Note that it is only for cases (i) and (ii) that we obtain a y_k that has the same qualitative behavior as the actual solution to the decay equation, namely, a monotonic decrease to zero. Quantitative agreement between $y(t)$ and y_k can be gotten by choosing h small, i.e., $0 < h \ll 1$.

The solution behaviors exhibited, in particular, by Figures 2.2.3(c), (d) and (e), we call numerical instabilities.

We now consider a forward Euler scheme with a symmetric form for the linear term; it is given by the expression

$$\frac{y_{k+1} - y_k}{h} = -\left(\frac{y_{k+1} + y_k}{2}\right). \tag{2.2.7}$$

Solving for y_{k+1} gives

$$y_{k+1} = \left(\frac{2 - h}{2 + h}\right)y_k. \tag{2.2.8}$$

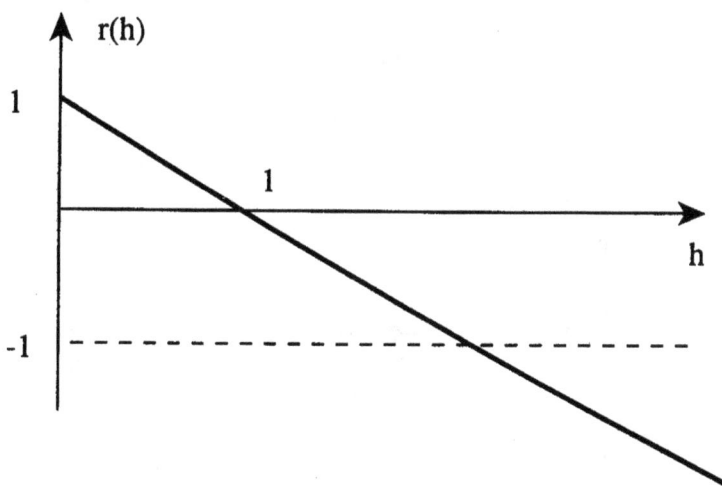

Figure 2.2.2. Plot of r(h) = 1 - h.

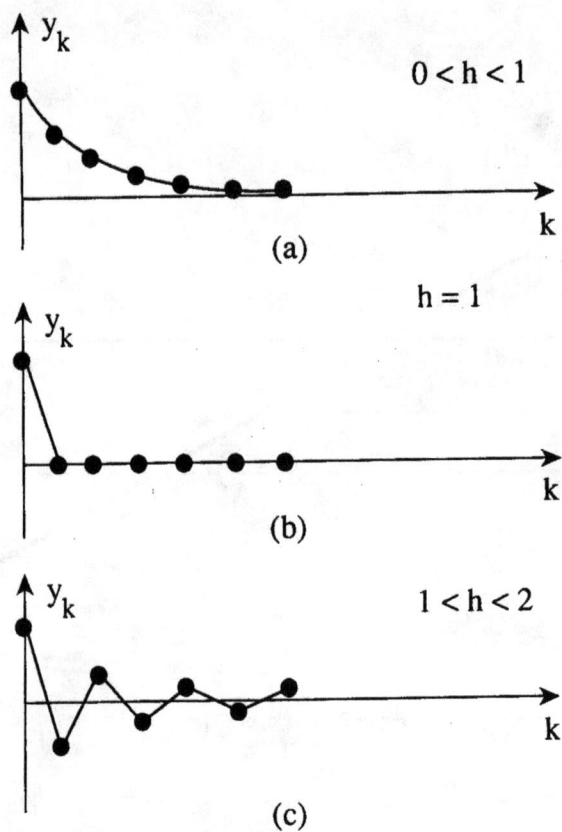

Figure 2.2.3. Plots of solutions to $y_{k+1} = (1-h)y_k$.

(d)

(e)

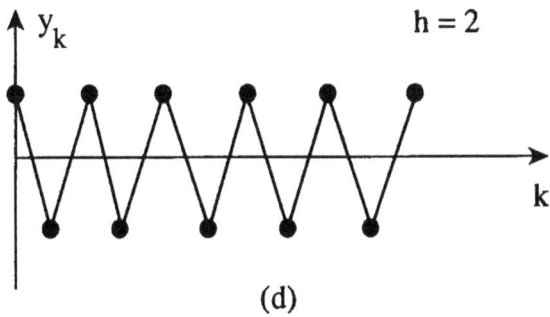

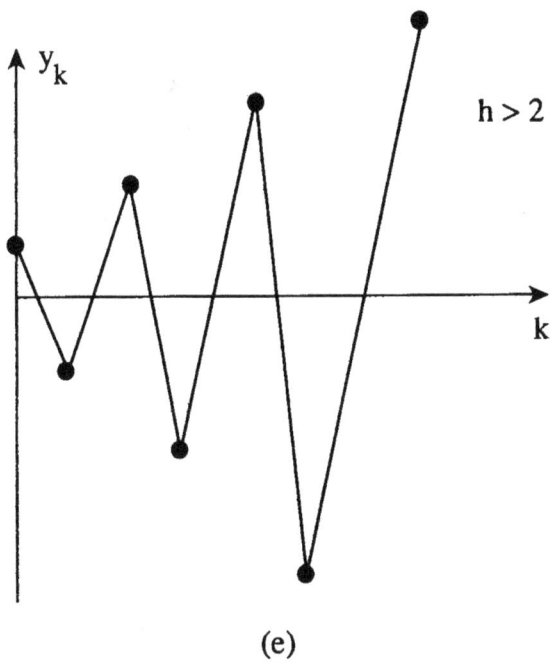

Figure 2.2.3. Plots of solutions to $y_{k+1} = (1-h)y_k$.

24

Again, this is a first-order, linear difference equation with constant coefficients. Its solution behavior is dependent on the value of

$$r(h) = \frac{2-h}{2+h}, \tag{2.2.9}$$

which is plotted in Figure 2.2.4. Since

$$|r(h)| < 1, \qquad 0 < h < \infty, \tag{2.2.10}$$

it follows that all solutions of Eq. (2.2.8)

$$y_k = y_0[r(h)]^k, \tag{2.2.11}$$

decrease to zero with $k \to \infty$. However, only for $0 < h < 2$ is the decrease monotonic. If $h > 2$, the solution is oscillatory with an amplitude that decreases exponentially. See Figure 2.2.5.

The hackward Euler scheme for the decay equation is

$$\frac{y_k - y_{k-1}}{h} = -y_k, \tag{2.2.12}$$

or

$$y_{k+1} = \left(\frac{1}{1+h}\right) y_k. \tag{2.2.13}$$

Since

$$0 < \frac{1}{1+h} < 1, \qquad 0 < h < \infty, \tag{2.2.14}$$

it follows that all the solutions of Eq. (2.2.13), i.e.,

$$y_k = y_0 \left(\frac{1}{1+h}\right)^k, \tag{2.2.15}$$

decrease (in magnitude) to zero monotonically for any step-size.

The central difference scheme is

$$\frac{y_{k+1} - y_{k-1}}{2h} = -y_k. \tag{2.2.16}$$

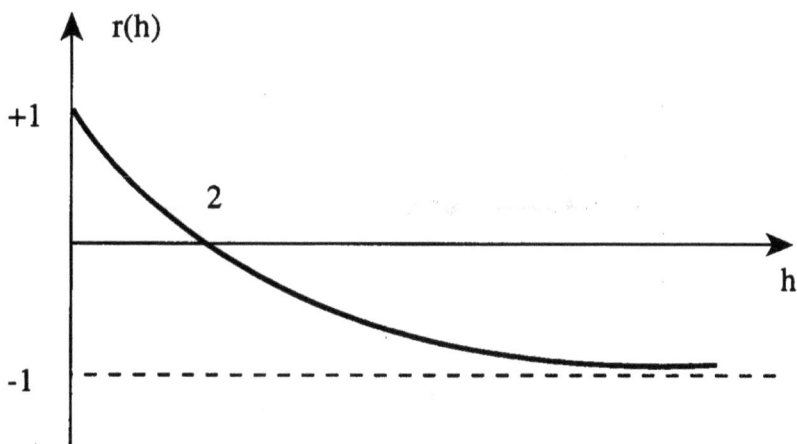

Figure 2.2.4. Plot of $r(h) = \dfrac{2-h}{2+h}$.

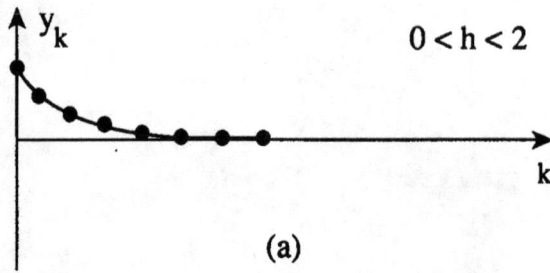

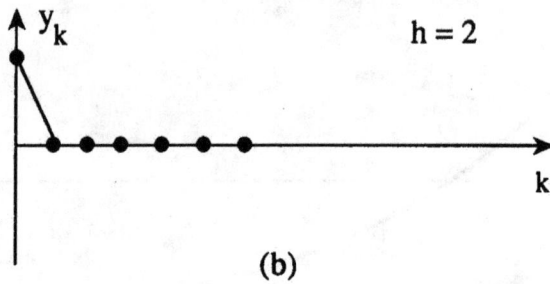

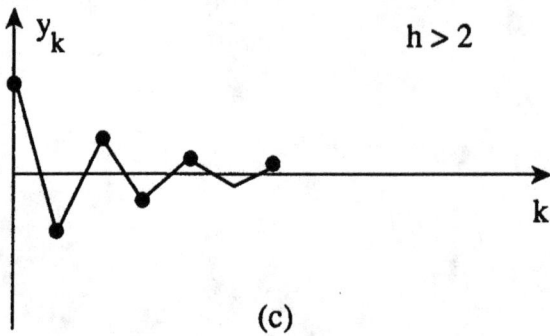

Figure 2.2.5. Plots of solutions to $y_{k+1} = \left(\dfrac{2-h}{2+h}\right) y_k$.

This is a second-order, linear difference equation

$$y_{k+2} + (2h)y_{k+1} - y_k = 0, \tag{2.2.17}$$

having constant coefficients. Its solution is

$$y_k = C_1(r_+)^k + C_2(r_-)^k, \tag{2.2.18}$$

where C_1 and C_2 are arbitrary constants, and (r_+, r_-) are solutions to the characteristic equation for Eq. (2.2.17)

$$r^2 + (2h)r - 1 = 0. \tag{2.2.19}$$

They are given by

$$r_+(h) = -h + \sqrt{1 + h^2}, \tag{2.2.20a}$$

$$r_-(h) = -h - \sqrt{1 + h^2}. \tag{2.2.20b}$$

An easy set of calculations shows the following to be true:

(i) $r_-(h) < -1$, $0 < h < \infty$;

(ii) $r_-(h) = -2h + O(\frac{1}{2h})$, for $h \to \infty$;

(iii) $0 < r_+(h) < 1$, $0 < h < \infty$;

(iv) $r_+(h) = \frac{1}{2h} + O(\frac{1}{h^3})$, for $h \to \infty$.

These facts lead to the conclusion that the second term on the right hand side of Eq. (2.2.18) oscillates with an amplitude that increases exponentially, while the first term decreases monotonically to zero. A typical solution trajectory is shown in Figure 2.2.6. Since this behavior holds for any step-size $h > 0$, we see that the central difference scheme has numerical instabilities regardless of the value of h.

In summary, the four discrete models of the decay equation only give the correct qualitative behavior for the numerical solution if the following conditions are satisfied for the step-size h;

(a) forward Euler: $0 < h < 1$;

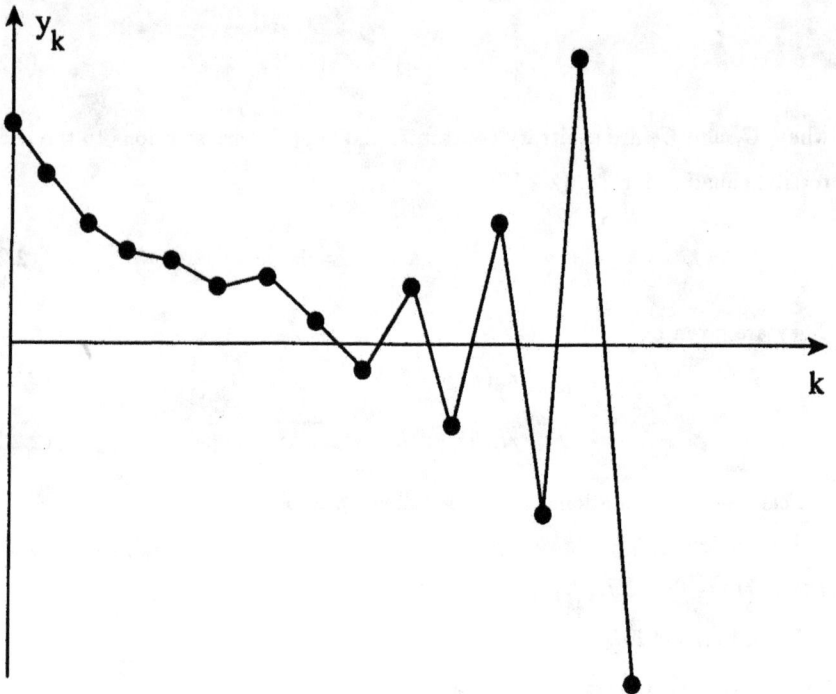

Figure 2.2.6. Plot of a typical solution to
$$\frac{y_{k+1} - y_{k-1}}{2h} = -y_k.$$

(b) forward Euler with symmetric linear term: $0 < h < 2$;

(c) backward Euler: $h > 0$;

(d) central difference: no value of h.

From these results, we conclude that the central difference scheme has numerical instabilities for all step-size values; the forward Euler schemes provide useful discrete models if limitations are placed on the step-size; and the backward Euler scheme can be used for any (positive) step-size. Except for the central difference scheme, the other three discrete models will give excellent quantitative numerical solutions if h is made small enough, i.e., $0 < h \ll 1$ [4].

2.3 Harmonic Oscillator

The harmonic oscillator differential equation

$$\frac{d^2y}{dt^2} + y = 0, \tag{2.3.1}$$

is characterized by the fact that all its solutions are periodic [5, 6]

$$y(t) = C_1 \cos t + C_2 \sin t = C e^{it} + C^* e^{-it}, \tag{2.3.2}$$

where C_1 and C_2 are arbitrary constants, and C is a complex valued constant.

The straightforward central difference scheme is

$$\frac{y_{k+1} - 2y_k + y_{k-1}}{h^2} + y_k = 0, \tag{2.3.3}$$

which can be rewritten to the form

$$y_{k+1} - (2 - h^2)y_k + y_{k-1} = 0. \tag{2.3.4}$$

This is a second-order, linear difference equation whose solution is

$$y_k = D_1(r_+)^k + D_2(r_-)^k, \tag{2.3.5}$$

where D_1 and D_2 are constants, and (r_+, r_-) are solutions to the characteristic equation

$$r^2 - 2\left(1 - \frac{h^2}{2}\right)r + 1 = 0. \tag{2.3.6}$$

Solving Eq. (2.3.6) gives

$$r_+(h) = \left(1 - \frac{h^2}{2}\right) + \left(\frac{h}{2}\right)\sqrt{h^2 - 4}, \tag{2.3.7a}$$

$$r_-(h) = \left(1 - \frac{h^2}{2}\right) - \left(\frac{h}{2}\right)\sqrt{h^2 - 4}. \tag{2.3.7b}$$

For $0 < h < 2$, $r_+(h)$ and $r_-(h)$ are complex valued with

$$r_+(h) = [r_-(h)]^* = \left(1 - \frac{h^2}{2}\right) + \left(\frac{ih}{2}\right)\sqrt{4 - h^2}. \tag{2.3.8}$$

They also have magnitude one since

$$|r_+(h)|^2 = |r_-(h)|^2 = \left(1 - \frac{h^2}{2}\right)^2 + \left(\frac{h^2}{4}\right)(4 - h^2) = 1. \tag{2.3.9}$$

Hence, for $0 < h < 2$, $r_+(h)$ and $r_-(h)$ have the representations

$$r_+(h) = [r_-(h)]^* = e^{i\phi(h)}, \tag{2.3.10}$$

$$\tan\phi(h) = \frac{(h/2)\sqrt{4 - h^2}}{\left(1 - \frac{h^2}{2}\right)}. \tag{2.3.11}$$

Consequently, the general solution to Eq. (2.3.4), for $0 < h < 2$, is

$$y_k = D_1 e^{i\phi(h)k} + D_2 e^{-i\phi(h)h}. \tag{2.3.12}$$

If $h = 2$, then

$$r_+(2) = r_-(2) = -1, \tag{2.3.13}$$

and the general solution is

$$y_k = (D_1 + D_2 k)(-1)^k. \tag{2.3.14}$$

For $h > 2$, $r_+(h)$ and $r_-(h)$ are both real and given by the expressions

$$r_+(h) = -\left(\frac{h^2}{2} - 1\right) + \left(\frac{h}{2}\right)\sqrt{h^2 - 4}, \qquad (2.3.15a)$$

$$r_-(h) = -\left(\frac{h^2}{2} - 1\right) - \left(\frac{h}{2}\right)\sqrt{h^2 - 4}, \qquad (2.3.15b)$$

It follows from Eq. (2.3.15b) that

$$r_-(h) < -1, \qquad h > 2, \qquad (2.3.16)$$

and from the characteristic Eq. (2.3.6) that

$$r_+(h)r_-(h) = 1. \qquad (2.3.17)$$

This implies that $r_+(h)$ must have a negative sign with a magnitude less than one, i.e.,

$$-1 < r_+(h) < 0, \qquad h > 2. \qquad (2.3.18)$$

Figure 2.3.1 gives the behavior of $r_+(h)$ and $r_-(h)$ as a function of h. Thus, for this case

$$y_k = \left[D_1|r_+(h)|^k + D_2|r_-(h)|^k\right](-1)^k, \qquad (2.3.19)$$

and we conclude that y_k will increase exponentially with an oscillating amplitude.

Putting all these results together, we observe that the straightforward central difference scheme has a solution with the same qualitative behavior as the harmonic oscillator differential equation only if the step-size is restricted to the interval $0 < h < 2$.

We now consider two central difference schemes for which the linear y term is modeled with a non-symmetric form. These discrete models are

$$\frac{y_{k+1} - 2y_k + y_{k-1}}{h^2} + y_{k-1} = 0, \qquad (2.3.20)$$

and

$$\frac{y_{k+1} - 2y_k + y_{k-1}}{h^2} + y_{k+1} = 0. \qquad (2.3.21)$$

32

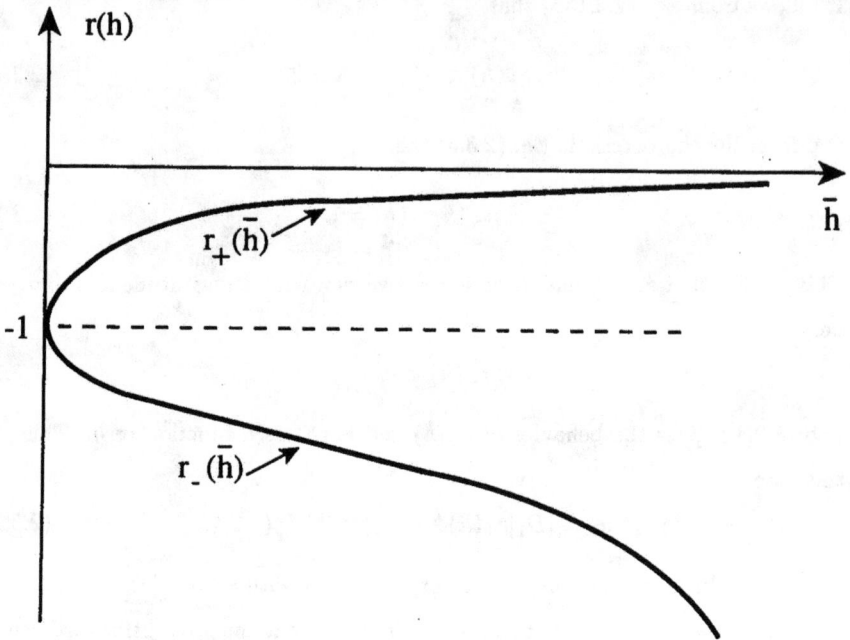

Figure 2.3.1. Plots of $r_+(\bar{h})$ and $r_-(\bar{h})$ from
Eqs. (2.3.15a) and (2.3.15b); $\bar{h} = h - 2$.

The characteristic equation for Eq. (2.3.20) is

$$r^2 - 2r + (1 + h^2) = 0, \tag{2.3.22}$$

with solutions

$$r_+(h) = [r_-(h)]^* = 1 + ih. \tag{2.3.23}$$

These can also be rewritten in the polar form

$$r_+(h) = \sqrt{1 + h^2}\, e^{i\phi(h)}, \tag{2.3.24}$$

$$\tan \phi(h) = h. \tag{2.3.25}$$

Note that the two roots are complex valued for all $h > 0$ and that they have magnitudes that are greater than one. As a consequence, all the solutions of this discrete model are oscillatory, but, they have an amplitude that increases exponentially.

Likewise, the characteristic equation for Eq. (2.3.21) is

$$r^2 - \left(\frac{2}{1 + h^2}\right) r + \left(\frac{1}{1 + h^2}\right) = 0. \tag{2.3.26}$$

Its solutions are again complex valued for all $h > 0$; they are

$$r_+(h) = [r_-(h)]^* = \frac{1 + ih}{1 + h^2} = \frac{1}{\sqrt{1 + h^2}}\, e^{i\phi(h)}, \tag{2.3.27}$$

where $\phi(h)$ is given by the relation of Eq. (2.3.25). Therefore, we conclude that, for $h > 0$, all the solutions of Eq. (2.3.21) are oscillatory with an amplitude that decreases exponentially.

We now examine the properties of a central difference scheme having a symmetric form for the linear term y. For the first example, we consider the following discrete model

$$\frac{y_{k+1} - 2y_k + y_{k-1}}{h^2} + \frac{y_{k+1} + y_{k-1}}{2} = 0. \tag{2.3.28}$$

Its characteristic equation is

$$r^2 - \left[\frac{2}{1+\frac{h^2}{2}}\right] r + 1 = 0, \tag{2.3.29}$$

with roots

$$r_{\pm}(h) = \left[\frac{1}{1+\frac{h^2}{2}}\right] \left[1 \pm ih\sqrt{1+\frac{h^2}{4}}\right]. \tag{2.3.30}$$

Note that

$$r_{+}(h) = [r_{-}(h)]^*, \qquad h > 0, \tag{2.3.31}$$

$$|r_{+}(h)| = |r_{-}(h)| = 1; \tag{2.3.32}$$

consequently,

$$r_{+} = r_{-}^* = e^{i\phi(h)}, \tag{2.3.33}$$

$$\tan\phi(h) = h\sqrt{1+\frac{h^2}{4}}. \tag{2.3.34}$$

Since

$$y_k = E(r_{+})^k + E^*(r_{+}^*)^k, \tag{2.3.35}$$

where E is an arbitrary complex-valued constant, we conclude that all solutions to this discrete model oscillate with constant amplitude for $h > 0$.

The second example has a completely symmetric discrete expression for the linear term; it is

$$\frac{y_{k+1} - 2y_k + y_{k-1}}{h^2} + \frac{y_{k+1} + y_k + y_{k-1}}{3} = 0. \tag{2.3.36}$$

The corresponding characteristic equation is

$$r^2 - 2\left[\frac{1-\frac{h^2}{6}}{1+\frac{h^2}{3}}\right] r + 1 = 0, \tag{2.3.37}$$

with the roots

$$r_{\pm}(h) = \left[\frac{1}{1+\frac{h^2}{3}}\right] \left\{\left(1-\frac{h^2}{6}\right) \pm ih\sqrt{1+\frac{h^2}{12}}\right\}. \tag{2.3.38}$$

These roots have the following properties:

$$r_+(h) = [r_-(h)]^*, \qquad h > 0, \tag{2.3.39}$$

$$|r_+(h)| = |r_-(h)| = 1, \qquad h > 0, \tag{2.3.40}$$

$$r_+(h) = [r_-(h)]^* = e^{i\phi(h)},$$

$$\tan \phi(h) = \frac{h\sqrt{1 + \frac{h^2}{12}}}{\left(1 - \frac{h^2}{6}\right)}. \tag{2.3.41}$$

We conclude that, for $h > 0$, all solutions of Eq. (2.3.36) are oscillatory with constant amplitude.

In summary, we have seen that only the use of a discrete representation for the linear y term that is centered about the grid point t_k will give a discrete model that has oscillations with constant amplitude. Non-centered schemes allow the amplitude of the oscillations to either increase or decrease. The straightforward central difference scheme has the correct oscillatory behavior if $0 < h < 2$, while the two "symmetric" forms for y give oscillatory behavior with constant amplitude for all $h > 0$.

2.4 Logistic Differential Equation

The Logistic differential equation is

$$\frac{dy}{dt} = y(1 - y). \tag{2.4.1}$$

Its exact solution can be obtained by the method of separation of variables which gives

$$y(t) = \frac{y_0}{y_0 + (1 - y_0)e^{-t}}, \tag{2.4.2}$$

where the initial condition is

$$y_0 = y(0). \tag{2.4.3}$$

Figure 2.4.1 illustrates the general nature of the various solution behaviors. If $y_0 > 0$, then all solutions monotonically approach the stable fixed-point at $y(t) = 1$. If $y_0 < 0$, then the solution at first decreases to $-\infty$ at the singular point

$$t = t^* = \text{Ln}\left[\frac{1 + |y_0|}{|y_0|}\right], \tag{2.4.4}$$

after which, for $t > t^*$, it decreases monotonically to the fixed-point at $y(t) = 1$. Note that $y(t) = 0$ is an unstable fixed-point.

Our first discrete model is constructed by using a central difference scheme for the derivative:

$$\frac{y_{k+1} - y_{k-1}}{2h} = y_k(1 - y_k). \tag{2.4.5}$$

Since Eq. (2.4.5) is a second-order difference equation, while Eq. (2.4.1) is a first-order differential equation, the value of $y_1 = y(h)$ must be determined by some procedure. We do this by use of the Euler result [7, 8, 9]

$$y_1 = y_0 + hy_0(1 - y_0). \tag{2.4.6}$$

A typical plot of the numerical solution to Eq. (2.4.5) is shown in Figure 2.4.2. This type of plot is obtained for any value of the step-size. An understanding of this result follows from a linear stability analysis of the two fixed points of Eq. (2.4.5).

First of all, note that Eq. (2.4.5) has two constant solutions or fixed-points. They are

$$y_k = \bar{y}^{(0)} = 0, \qquad y_k = \bar{y}^{(1)} = 1. \tag{2.4.7}$$

To investigate the stability of $y_k = \bar{y}^{(0)}$, we set

$$y_k = \bar{y}^{(0)} + \epsilon_k, \qquad |\epsilon_k| \ll 1, \tag{2.4.8}$$

substitute this result into Eq. (2.4.5) and neglect all but the linear terms. Doing this gives

$$\frac{\epsilon_{k+1} - \epsilon_{k-1}}{2h} = \epsilon_k. \tag{2.4.9}$$

(a)

(b)

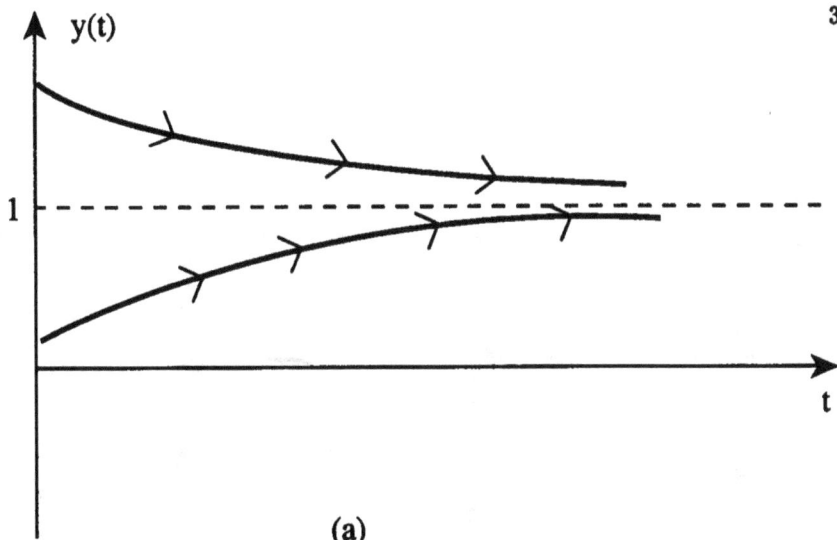

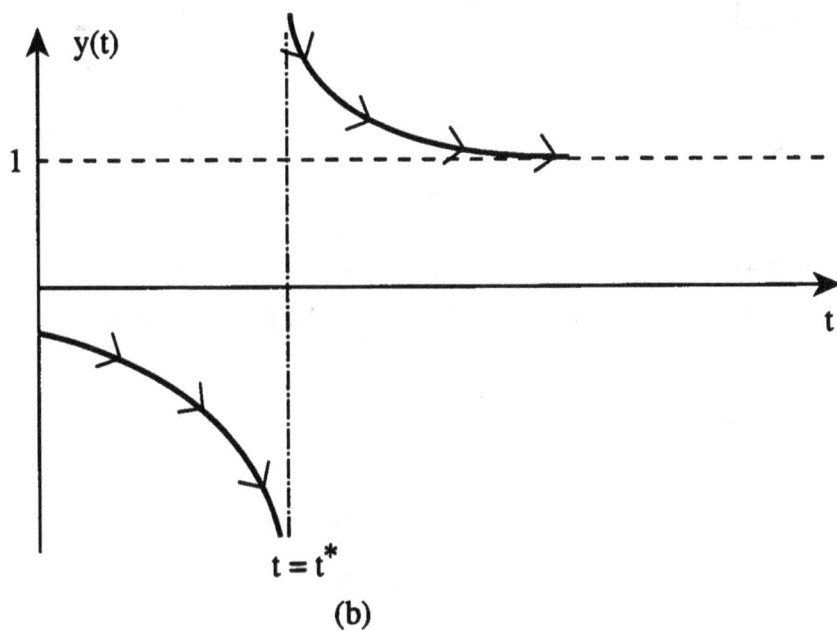

Figure 2.4.1. Solutions of the logistic differential
equation. (a) $y_0 > 0$. (b) $y_0 < 0$.

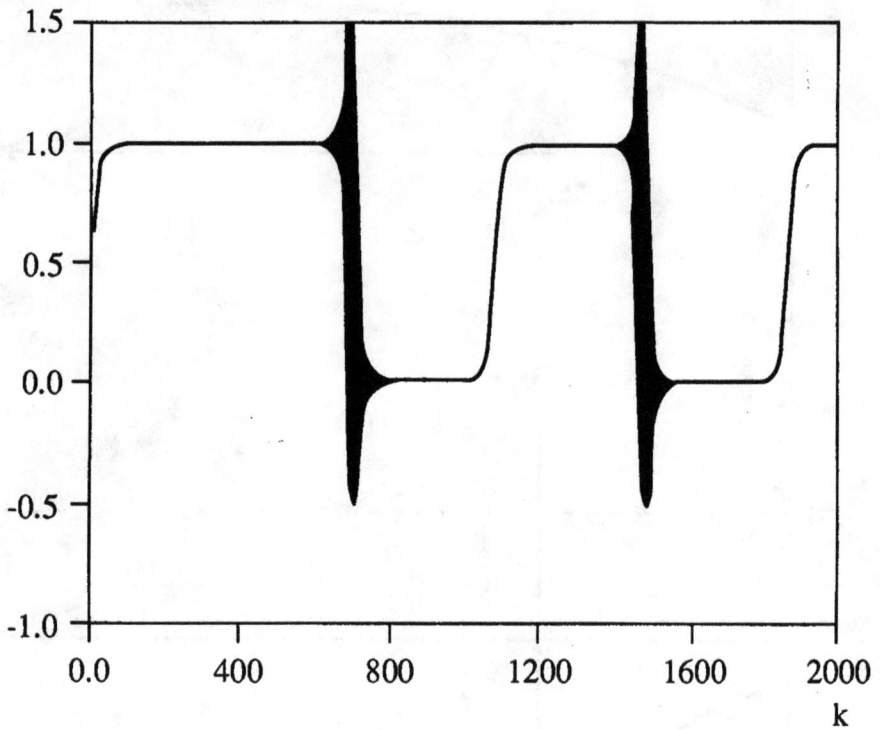

Figure 2.4.2. Typical plot for a central difference scheme model of the logistic differential equation: $y_0 = 0.5$, $h = 0.1$.

$$\frac{y_{k+1} - y_{k-1}}{2h} = y_k(1 - y_k).$$

The solution to this second-order difference equation is

$$\epsilon_k = A(r_+)^k + B(r_-)^k, \qquad (2.4.10)$$

where A and B are arbitrary, but, small constants; and

$$r_\pm(h) = h \pm \sqrt{1 + h^2}. \qquad (2.4.11)$$

From Eq. (2.4.11), it can be concluded that the first term on the right-side of Eq. (2.4.10) is exponentially increasing, while the second term oscillates with an exponentially decreasing amplitude.

A small perturbation to the fixed-point at $\bar{y}^{(1)} = 1$ can be represented as

$$y_k = \bar{y}^{(1)} + \eta_k, \qquad |\eta_k| \ll 1. \qquad (2.4.12)$$

The linear perturbation equation for η_k is

$$\frac{\eta_{k+1} - \eta_{k-1}}{2h} = -\eta_k, \qquad (2.4.13)$$

whose solution is

$$\eta_k = C(S_+)^k + D(S_-)^k, \qquad (2.4.14)$$

where C and D are small arbitrary constants, and

$$S_\pm(h) = -h \pm \sqrt{1 + h^2}. \qquad (2.4.15)$$

Thus, the first term on the right-side of Eq. (2.4.14) exponentially decreases, while the second term oscillates with an exponentially increasing amplitude.

Putting these results together, it follows that the central difference scheme has exactly the same two fixed-points as the Logistic differential equation. However, while $y(t) = 0$ is (linearly) stable and $y(t) = 1$ is (linearly) unstable for the differential equation, both fixed-points are linearly unstable for the central difference scheme. The results of the linear stability analysis, as given in Eqs. (2.4.10) and

(2.4.14), explain what is shown by Figure 2.4.2. For initial value y_0, such that $0 < y_0 < 1$, the values of y_k increase and exponentially approach the fixed-point $\bar{y}^{(1)} = 1$; y_k then begins to oscillate with an exponentially increasing amplitude about $\bar{y}^{(1)} = 1$ until it reaches the neighborhood of the fixed-point $\bar{y}^{(0)} = 0$. After an initial exponential decrease to $\bar{y}^{(0)} = 0$, the y_k value then begin their increase back to the fixed-point at $\bar{y}^{(1)} = 0$.

It has been shown by Yamaguti and Ushiki [10] and by Ushiki [11] that the central difference scheme allows for the existence of chaotic orbits for all positive time-steps for the Logistic differential equation. Additional work on this problem has been done by other researchers including Sanz-Serna [12] and Mickens [13]. The major conclusion is that the use of the central difference scheme

$$\frac{y_{k+1} - y_{k-1}}{2h} = f(y_k), \qquad (2.4.16)$$

for the scalar first-order differential equation

$$\frac{dy}{dt} = f(y) \qquad (2.4.17)$$

forces all the fixed-points to become unstable [13]. Consequently, the central difference discrete derivative should never be used for this class of ordinary differential equation.

However, before leaving the use of the central difference scheme, let us consider the following discrete model for the Logistic equation:

$$\frac{y_{k+1} - y_{k-1}}{2h} = y_{k-1}(1 - y_{k+1}). \qquad (2.4.18)$$

Our major reason for studying this model is that an exact analytic solution exists for Eq. (2.4.18). Observe that the function

$$f(y) = y(1 - y) \qquad (2.4.19)$$

is modeled locally on the lattice in Eq. (2.4.5), while it is modeled nonlocally in Eq. (2.4.18), i.e., at lattice points $k - 1$ and $k + 1$.

The substitution

$$y_k = \frac{1}{x_k},$$
(2.4.20)

transforms Eq. (2.4.18) to the expression

$$x_{k+1} - \left(\frac{1}{1+2h}\right)x_{k-1} = \frac{2h}{1+2h}.$$
(2.4.21)

Note that Eq. (2.4.18) is a nonlinear, second-order difference equation, while Eq. (2.4.21) is a linear, inhomogeneous equation with constant coefficients. Solving Eq. (2.4.21) gives the general solution

$$x_k = 1 + [A + B(-1)^k](1 + 2h)^{-k/2},$$
(2.4.22)

where A and B are arbitrary constants. Therefore, y_k is

$$y_k = \frac{1}{1 + [A + B(-1)^k](1 + 2h)^{-k/2}}.$$
(2.4.23)

For y_0 such that $0 < y_0 < 1$, and y_1 selected such that $y_1 = y_0 + hy_0(1 - y_0)$, the solutions to Eq. (2.4.23) have the structure indicated in Figure 2.4.3. Observe that the numerical solution has the general properties of the solution to the Logistic differential equation, see Figure 2.4.1, except that small oscillations occur about the smooth solution.

The direct forward Euler discrete model for the Logistic differential equation is

$$\frac{y_{k+1} - y_k}{h} = y_k(1 - y_k).$$
(2.4.24)

This first-order difference equation has two constant solutions or fixed-points at $\bar{y}^{(0)} = 0$ and $\bar{y}^{(1)} = 1$. Perturbations about these fixed-points, i.e.,

$$y_k = \bar{y}^{(0)} + \epsilon_k = \epsilon_k, \qquad |\epsilon_k| \ll 1,$$
(2.4.25)

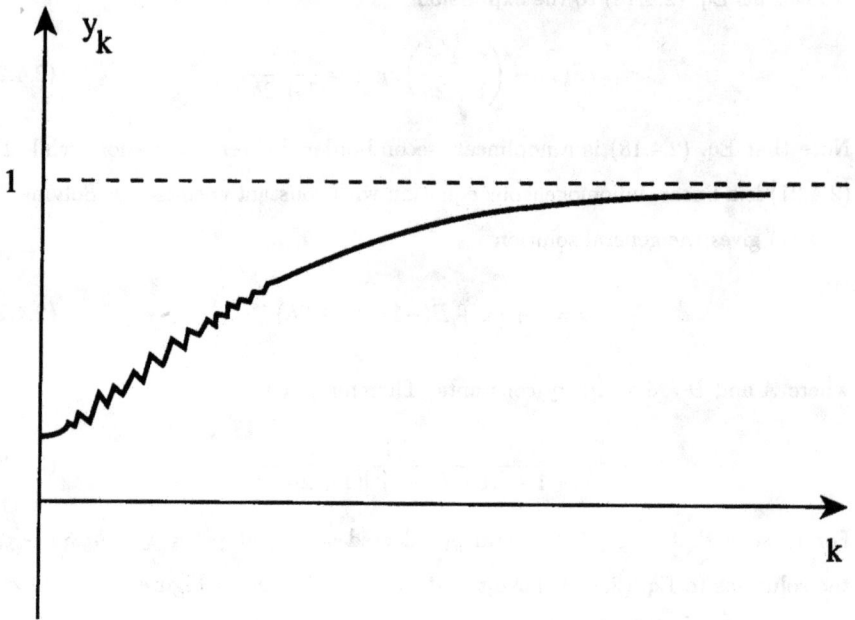

Figure 2.4.3. A trajectory for the central difference scheme

$$\frac{y_{k+1} - y_{k-1}}{2h} = y_{k-1}(1 - y_{k+1}).$$

$$y_k = \bar{y}^{(1)} + \eta_k = 1 + \eta_k, \qquad |\eta_k| \ll 1, \qquad (2.4.26)$$

give the following solutions for ϵ_k and η_k:

$$\epsilon_k = \epsilon_0(1 + h)^k, \qquad (2.4.27)$$

$$\eta_k = \eta_0(1 - h)^k. \qquad (2.4.28)$$

The expression for ϵ_k shows that $\bar{y}^{(0)}$ is unstable for all $h > 0$; thus, this discrete scheme has the same linear stability property as the differential equation for all $h > 0$. However, the linear stability properties of the fixed-point $\bar{y}^{(1)}$ depend on the value of the step-size. For example:

(i) $0 < h < 1 : \bar{y}^{(1)}$ is linearly stable; perturbations decrease exponentially.

(ii) $1 < h < 2 : \bar{y}^{(1)}$ is linearly stable; however, the perturbations decrease exponentially with an oscillating amplitude.

(iii) $h > 2 : \bar{y}^{(1)}$ is linearly unstable; the perturbations oscillate with an exponentially increasing amplitude.

Our conclusion is that the forward Euler scheme gives the correct linear stability properties only if $0 < h < 1$. For this interval of step-size values, the qualitative properties of the solutions to the differential and difference equations are the same. Consequently, for $0 < h < 1$, there are no numerical instabilities.

Figure 2.4.4 presents three numerical solutions for the forward Euler scheme given by Eq. (2.4.24). In all three cases the initial condition is $y_0 = 0.5$. The step-sizes are $h = 0.01$, 1.5 and 2.5.

Finally, it should be stated that the change of variables

$$z_k = \left(\frac{h}{1+h}\right)y_k, \qquad \lambda = 1 + h, \qquad (2.4.29)$$

when substituted into Eq. (2.4.24) gives the famous Logistic difference equation [7, 14, 15]

$$z_{k+1} = \lambda z_k(1 - z_k). \qquad (2.4.30)$$

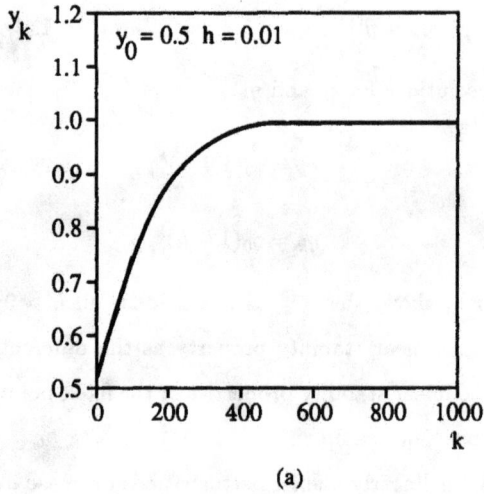

(a)

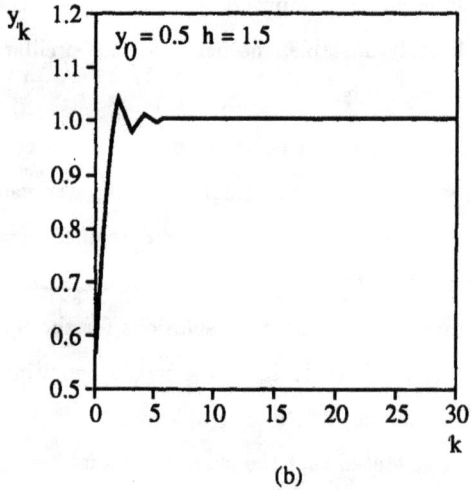

(b)

Figure 2.4.4. The forward Euler scheme

$$\frac{y_{k+1} - y_k}{h} = y_k(1 - y_k).$$

(a) $y_0 = 0.5$, $h = 0.01$. (b) $y_0 = 0.5$, $h = 1.5$.

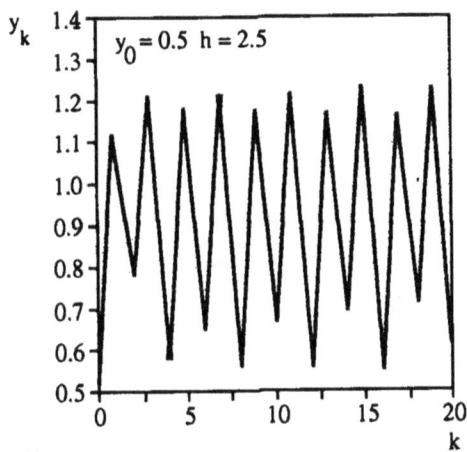

(c)

Figure 2.4.4. The forward Euler scheme

$$\frac{y_{k+1} - y_k}{h} = y_k(1 - y_k).$$

(c) $y_0 = 0.5$, $h = 2.5$.

Depending on the value of the parameter λ, this equation has a host of solutions with various periods, as well as chaotic solutions [16, 17].

Our next model of the Logistic differential equation is constructed by using a forward Euler for the first-derivative and a nonlocal expression for the function $f(y) = y(1 - y)$. This model is

$$\frac{y_{k+1} - y_k}{h} = y_k(1 - y_{k+1}). \tag{2.4.31}$$

This first-order, nonlinear difference equation can be solved exactly by using the variable change

$$y_k = \frac{1}{x_k}, \tag{2.4.32}$$

to obtain

$$x_{k+1} - \left(\frac{1}{1+h}\right)x_k = \frac{h}{1+h}, \tag{2.4.33}$$

whose general solution is

$$x_k = 1 + A(1 + h)^{-k}, \tag{2.4.34}$$

where A is an arbitrary constant. Imposing the initial condition

$$x_0 = \frac{1}{y_0}, \tag{2.4.35}$$

gives

$$A = \frac{1 - y_0}{y_0}, \tag{2.4.36}$$

and

$$y_k = \frac{y_0}{y_0 + (1 - y_0)(1 + h)^{-k}}. \tag{2.4.37}$$

Examination of Eq. (2.4.37) shows that, for $h > 0$, its qualitative properties are the same as the corresponding exact solution to the Logistic differential equation, namely, Eq. (2.4.2). Hence, the forward Euler, nonlocal discrete model has no numerical instabilities for any step-size. Figure 2.4.5 gives numerical solutions using

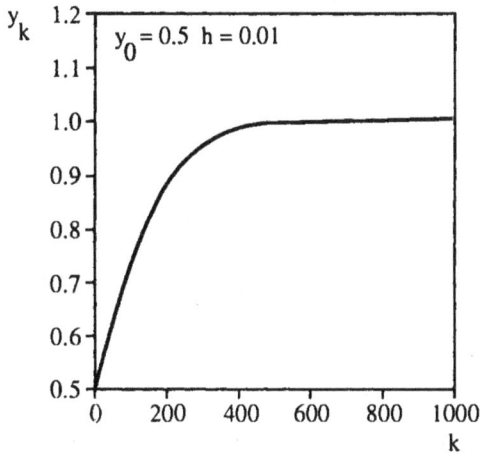

(a)

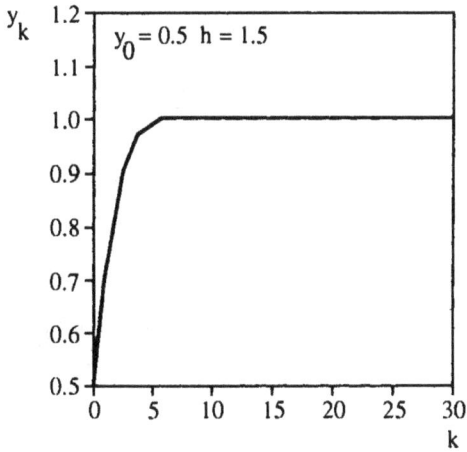

(b)

Figure 2.4.5. Numerical solutions of

$$\frac{y_{k+1} - y_k}{h} = y_k (1 - y_{k+1}).$$

(a) $y_0 = 0.5$, h = 0.01. (b) $y_0 = 0.5$, h = 1.5.

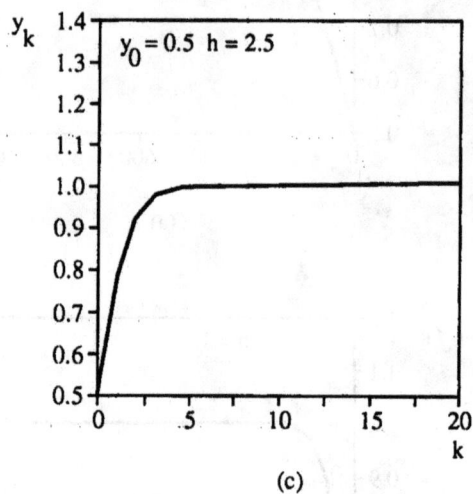

Figure 2.4.5. Numerical solutions of

$$\frac{y_{k+1} - y_k}{h} = y_k (1 - y_{k+1}).$$

(c) $y_0 = 0.5$, $h = 2.5$.

Eq. (2.4.31) for three step-sizes. Note that Eq. (2.4.31) can be written in explicit form

$$y_{k+1} = \frac{(1+h)y_k}{1+hy_k}. \tag{2.4.38}$$

Our last discrete model for the Logistic differential equation is based on a second-order Runge-Kutta method [8, 9]. This technique gives for the first-order scalar equation

$$\frac{dy}{dt} = f(y), \tag{2.4.39}$$

the discrete result

$$\frac{y_{k+1} - y_k}{h} = \frac{f(y_k) + f[y_k + hf(y_k)]}{2}. \tag{2.4.40}$$

Applying this to the Logistic equation, where $f(y) = y(1-y)$, gives

$$y_{k+1} = \left[1 + \frac{(2+h)h}{2}\right]y_k - \left[\frac{(2+3h+h^2)h}{2}\right]y_k^2 + (1+h)h^2y_k^3 - \left(\frac{h^3}{2}\right)y_k^4. \tag{2.4.41}$$

This first-order, nonlinear difference equation has four fixed-points. They are located at

$$\bar{y}^{(0)} = 0, \qquad \bar{y}^{(1)} = 1, \tag{2.4.42}$$

$$\bar{y}^{(2,3)} = \left(\frac{1}{2h}\right)\left[(2+h) \pm \sqrt{h^2 - 4}\right]. \tag{2.4.43}$$

The first two fixed-points, $\bar{y}^{(0)}$ and $\bar{y}^{(1)}$, correspond to the two fixed-points of the Logistic differential equation. The other two fixed-points, $\bar{y}^{(2)}$ and $\bar{y}^{(3)}$, are spurious fixed-points and are introduced by the second-order Runge-Kutta method. Note that for $h \leq 2$, the fixed-points $\bar{y}^{(2)}$ and $\bar{y}^{(3)}$ are complex conjugates of each other; while for $h \geq 2$, all fixed-points are real. Figure 2.4.6 gives a plot of all the fixed-points as a function of the step-size h.

For $0 < h < 2$, there are only two real fixed-points, namely, $\bar{y}^{(0)} = 0$ and $\bar{y}^{(1)} = 1$. The first is linearly unstable and the second is linearly stable. All numerical solutions of Eq. (2.4.41), with $y_0 > 0$, thus approach $\bar{y}^{(1)}$ as $k \to \infty$. However, for $h > 2$, there exists four real fixed-points. Their order and linear

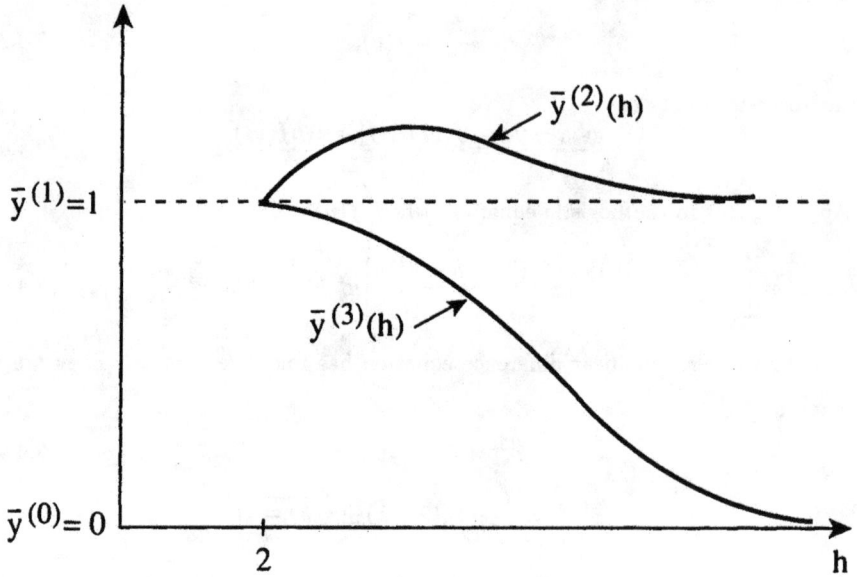

Figure 2.4.6. Plot of the fixed-points of the 2nd-order Runge-Kutta method for the logistic differential equation. Only the spurious fixed-points depend on h.

stability properties are indicated below where U and S, respectively, mean linearly unstable and linearly stable:

$$\bar{y}^{(0)} < \bar{y}^{(3)}(h) < \bar{y}^{(1)} < \bar{y}^{(2)}(h)$$

$$U \qquad S \qquad U \qquad S.$$

These results and Eq. (2.4.43) predict that at a step-size of $h = 2.5$, if the initial value y_0 is selected so that $0 < y_0 < 1$, then the numerical solution of Eq. (2.4.41) will converge to the value 0.6. The validity of this prediction is shown in Figure 2.4.7(c). This figure also gives numerical solutions for several other step-sizes.

The application of the second-order Runge-Kutta method illustrates the generation of numerical instabilities that arise from the creation of additional spurious fixed-points.

Comparing the five finite-difference schemes that were used to model the Logistic differential equation, the nonlocal forward Euler method clearly gave the best results. For all values of the step-size it has solutions that are in qualitative agreement with the corresponding solutions of the differential equation. The other discrete models had, for certain values of step-size, numerical instabilities.

2.5 Unidirectional Wave Equation

The one-way or unidirectional wave equation is [10]

$$u_t + u_x = 0, \qquad u(x,0) = f(x), \tag{2.5.1}$$

where the initial profile function $f(x)$ is assumed to have a first derivative. The solution to the initial value problem of Eq. (2.5.1) is

$$u(x,t) = f(x - t). \tag{2.5.2}$$

This represents a waveform moving to the right with unit velocity.

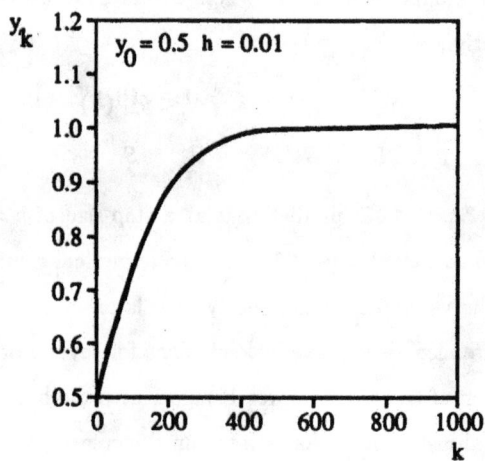

(a)

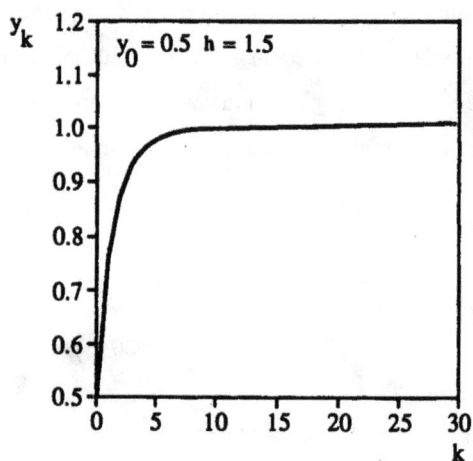

(b)

Figure 2.4.7. Numerical integration of the logistic equation
by a 2nd-order Runge-Kutta method.
(a) $y_0 = 0.5$, $h = 0.01$. (b) $y_0 = 0.5$, $h = 1.5$.

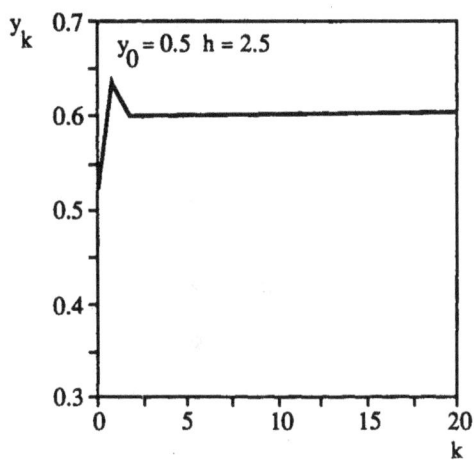

(c)

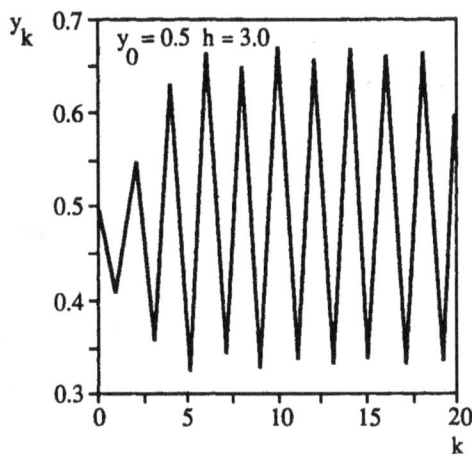

(d)

Figure 2.4.7. Numerical integration of the logistic equation
by a 2nd-order Runge-Kutta method.
(c) $y_0 = 0.5$, $h = 2.5$. (d) $y_0 = 0.5$, $h = 3.0$.

A discrete model for the unidirectional wave equation that uses forward Euler expressions for both the time and space derivatives is

$$\frac{u_m^{k+1} - u_m^k}{\Delta t} + \frac{u_{m+1}^k - u_m^k}{\Delta x} = 0, \qquad (2.5.3)$$

where

$$t_k = (\Delta t)k, \qquad x_m = (\Delta x)m, \qquad (2.5.4)$$

are the discrete time and space variables. Define β to be

$$\beta = \frac{\Delta t}{\Delta x}. \qquad (2.5.5)$$

Using this definition, Eq. (2.5.3) takes the form

$$u_m^{k+1} + \beta u_{m+1}^k - (1+\beta)u_m^k = 0. \qquad (2.5.6)$$

The method of separation of variables can be used to obtain particular solutions to Eq. (2.5.6). Assume that u_m^k can be written as [7]

$$u_m^k = C^k D_m. \qquad (2.5.7)$$

(Note that C^k is a function of the discrete variable k and does not mean "C" raised to the k-th power. The same statement applies to D_m.) Substitution of this result into Eq. (2.5.6) gives

$$C^{k+1} D_m + C^k [\beta D_{m+1} - (1+\beta)D_m] = 0, \qquad (2.5.8)$$

and

$$\frac{C^{k+1}}{C^k} + \frac{\beta D_{m+1} - (1+\beta)D_m}{D_m} = 0. \qquad (2.5.9)$$

Since the first term depends only on k, while the second term depends only on m, each term must be constant. Denoting the "separation constant" by α, we obtain

$$C^{k+1} = \alpha C^k, \qquad (2.5.10)$$

$$\beta D_{m+1} - (1 + \beta)D_m = -\alpha D_m. \qquad (2.5.11)$$

These equations have the respective solutions

$$C^k = A(\alpha)\alpha^k, \qquad (2.5.12)$$

$$D_m = B(\alpha)\left(\frac{1 + \alpha + \beta}{\beta}\right)^m, \qquad (2.5.13)$$

where $A(\alpha)$ and $B(\alpha)$ are "constants." Therefore, a particular solution to Eq. (2.5.6) is

$$u_m^k(\alpha, \beta) = E(\alpha)\alpha^k\left(\frac{1 + \alpha + \beta}{\beta}\right)^m, \qquad (2.5.14)$$

$$E(\alpha) = A(\alpha)B(\alpha). \qquad (2.5.15)$$

The Eq. (2.5.14) can also be written in the form

$$u_m^k(\alpha, \beta) = E(\alpha)\alpha^{(k-m)}\left[\frac{(1 + \alpha + \beta)\alpha}{\beta}\right]^m. \qquad (2.5.16)$$

If β is chosen to be

$$\beta = 1, \qquad (2.5.17)$$

and if a sum/integral is done over α, then the following general solution is obtained [7]

$$
\begin{aligned}
u_m^k &= \sum_{\alpha}\!\!\!\!\!\int u_m^k(\alpha, 1) \\
&= \sum_{\alpha}\!\!\!\!\!\int \bar{\alpha}^{(t_k - x_m)}(2 + \alpha)^m, \qquad \bar{\alpha} = \alpha^{1/\Delta x}, \\
&\neq g(x_m - t_k),
\end{aligned}
\qquad (2.5.18)
$$

where $g(z)$ is an arbitrary function of z. Our general conclusion is that the finite-difference scheme, of Eq. (2.5.3), does not have solutions that correspond exactly to those of the unidirectional wave equation.

For a second model, let us replace the time and space derivatives by, respectively, forward Euler and central difference expressions. Doing this gives

$$\frac{u_m^{k+1} - u_m^k}{\Delta t} + \frac{u_{m+1}^k - u_{m-1}^k}{\Delta x} = 0, \qquad (2.5.19)$$

and, upon rearranging, the equation

$$u_m^{k+1} + \beta u_{m+1}^k - u_m^k - \beta u_{m-1}^k = 0. \qquad (2.5.20)$$

Assuming a particular solution, $u_m^k = C^k D_m$, we obtain

$$C^{k+1} D_m - C^k [D_{m+1} - D_m - \beta D_{m-1}] = 0. \qquad (2.5.21)$$

Let ζ be the separation constant. The equations for C^k and D_m are

$$C^{k+1} = \zeta C^k, \qquad (2.5.22)$$

$$\beta D_{m+1} + (\zeta - 1) D_m - \beta D_{m-1} = 0. \qquad (2.5.23)$$

These difference equations have the solutions

$$C^k = A(\zeta) \zeta^k \qquad (2.5.24)$$

$$D_m = B_1(\zeta)[r_+(\beta,\zeta)]^k + B_2(\zeta)[r_-(\beta,\zeta)]^k, \qquad (2.5.25)$$

where $r_+(\beta,\zeta)$ and $r_-(\beta,\zeta)$ are roots to the characteristic equation [7]

$$\beta r^2 + (\zeta - 1) r - \beta = 0. \qquad (2.5.26)$$

Therefore,

$$u_m^k(\beta,\zeta) = \{H_1(\zeta)[r_+(\beta,\zeta)]^m + H_2(\zeta)[r_-(\beta,\zeta)]^m \zeta^k, \qquad (2.5.27)$$

and a general solution is

$$u_m^k(\beta) = \sum_\zeta u_m^k(\beta,\zeta) \neq g(x_m - t_k). \qquad (2.5.28)$$

Again, we conclude that Eq. (2.5.19) does not provide a good discrete model for the unidirectional wave equation.

Finally, consider a discrete model for which the time and space derivatives are given, respectively, by forward and backward Euler expressions. For this case, we have

$$\frac{u_m^{k+1} - u_m^k}{\Delta t} + \frac{u_m^k - u_{m-1}^k}{\Delta x} = 0 \tag{2.5.29}$$

and

$$u_m^{k+1} + (\beta - 1)u_m^k - \beta u_{m-1}^k = 0. \tag{2.5.30}$$

The separation of variables equations are

$$C^{k+1} = \gamma C^k, \tag{2.5.31}$$

$$(\beta + \gamma - 1)D_m - \beta D_{m-1} = 0, \tag{2.5.32}$$

where γ is the separation constant. The solutions to these equations are

$$C^k = A(\gamma)\gamma^k, \tag{2.5.33}$$

$$D_m = B(\gamma)\left(\frac{1}{\beta + \gamma - 1}\right)^m. \tag{2.5.34}$$

With $G(\gamma) = A(\gamma)B(\gamma)$, the general solution is

$$u_m^k(\beta) = \sum_\gamma G(\gamma)\gamma^k(\beta + \gamma - 1)^{-m}. \tag{2.5.35}$$

Note that for general β, we have

$$u_m^k(\beta) \neq g(x_m - t_k). \tag{2.5.36}$$

However, for $\beta = 1$ or $\Delta t = \Delta x$, Eq. (2.5.35) becomes

$$u_m^k(1) = \sum_\gamma G(\gamma)\gamma^{(k-m)} = g(x_m - t_k). \tag{2.5.37}$$

Consequently, if $\beta = 1$, the discrete model of Eq. (2.5.29) has solutions that are exactly equal to the solution of the unidirectional wave equation on the computational lattice. Under this condition, Eq. (2.5.29) reduces to the simpler form

$$u_m^{k+1} = u_{m-1}^k, \qquad \Delta t = \Delta x. \tag{2.5.38}$$

2.6 Burgers' Equation

The full Burgers' equation is [18, 19]

$$u_t + uu_x = \epsilon u_{xx}, \tag{2.6.1}$$

where ϵ is related to the reciprocal of the Reynolds number [18]. For the present study, we consider the case for which $\epsilon = 0$, i.e.,

$$u_t + uu_x = 0. \tag{2.6.2}$$

This first-order, nonlinear partial differential equation has no exact explicit solution for the initial value problem $u(x,0) = f(x)$ where $f(z)$ has a first derivative [18]. However, it does have a particular solution that can be obtained by the method of separation of variables. Assume

$$u(x,t) = C(t)D(x), \tag{2.6.3}$$

and substitute this into Eq. (2.6.2) to obtain

$$\frac{dC}{dt}D + CDC\frac{dD}{dx} = 0, \tag{2.6.4}$$

or

$$\frac{1}{C^2}\frac{dC}{dt} + \frac{dD}{dx} = 0. \tag{2.6.5}$$

Denoting the separation constant by a, these ordinary differential equations have the solutions

$$C(t) = \frac{1}{at+d}, \tag{2.6.6}$$

$$D(x) = ax + b, \tag{2.6.7}$$

where b and d are arbitrary integration constants. Hence, a particular, rational solution of the Burgers' equation is

$$u(x,t) = \frac{ax + b}{at + d}. \tag{2.6.8}$$

Now consider a discrete model of the Burgers' equation that uses forward Eulers for both the time and space derivatives:

$$\frac{u_m^{k+1} - u_m^k}{\Delta t} + u_m^k \left(\frac{u_{m+1}^k - u_m^k}{\Delta x} \right) = 0. \tag{2.6.9}$$

For $\beta = \Delta t / \Delta x$, this equation takes the form

$$u_m^{k+1} - u_m^k + \beta u_m^k u_{m+1}^k - \beta (u_m^k)^2 = 0. \tag{2.6.10}$$

Let $u_m^k = C^k D_m$, then the equations satisfied by C^k and D_m are

$$C^{k+1} = C^k - \alpha (C^k)^2, \tag{2.6.11}$$

$$D_{m+1} - D_m = \frac{\alpha}{\beta}, \tag{2.6.12}$$

where α is the separation constant. The general solution to Eq. (2.6.12) is

$$D_m = \left(\frac{\alpha}{\beta} \right) m + b_1, \tag{2.6.13}$$

where b_1 is an arbitrary constant. If we now make the identifications

$$\frac{\alpha}{\beta} = a(\Delta x), \qquad b_1 = b, \tag{2.6.14}$$

then Eq. (2.6.13) becomes

$$D_m = ax_m + b, \tag{2.6.15}$$

which is the discrete version of Eq. (2.6.7). However, Eq. (2.6.11) is the Logistic difference equation for which no general solution exists in terms of a finite sum of elementary functions [1, 7]. Therefore, the discrete version of Eq. (2.6.6) is not a solution of Eq. (2.6.11). Our conclusion is that Eq. (2.6.9) will have solutions that do not correspond to any solution of the Burgers' partial differential equation; consequently, this scheme has numerical instabilities.

Similar results are obtained for the discrete model

$$\frac{u_m^{k+1} - u_m^k}{\Delta t} + u_m^k \left(\frac{u_m^k - u_{m-1}^k}{\Delta x} \right) = 0 \qquad (2.6.16)$$

for which the separation of variables equations are

$$C^{k+1} = C^k - \alpha\beta(C^k)^2, \qquad (2.6.17)$$

$$D_{m+1} - D_m = \frac{\alpha}{\beta}. \qquad (2.6.18)$$

2.7 Summary

What have we learned from the study of various discrete models of several linear and nonlinear differential equations? The results stated below are based not only on those equations investigated in this chapter, but also on other differential equations and their associated finite-difference models [20–22].

First, if the order of the finite-difference scheme is greater than the order of the differential equation, then numerical instabilities will certainly occur for all step-sizes. This type of behavior is illustrated by the use of the central difference scheme for the first derivative in both the decay and Logistic equations. Mathematically, this type of instability occurs because the higher-order difference equation has a larger set of general solutions than the corresponding differential equation. For example, the linear decay equation has but one solution. However, a discrete model that uses the central difference scheme has two linearly independent solutions since it is of second-order.

Second, most discrete models require restrictions on the step-size to ensure that numerical instabilities do not occur. All forward Euler type schemes and their generalizations, such as Runge-Kutte methods, have this property.

Third, for many ordinary differential equations, a linear stability analysis of the fixed-points allows a determination of when numerical instabilities occur.

Fourth, the use of nonlocal representations of non-derivative terms can often eliminate numerical instabilities, as was the case for the Logistic differential equation with a forward Euler discrete derivative. In some instances, for example, in the application of the central difference scheme to the Logistic equation, a nonlocal model gave solutions that followed rather closely the trajectories of the solution to the differential equation except for small oscillations.

Fifth, for discrete models of partial differential equations, the use of forward or backward Euler schemes for the first-derivatives can have a significant impact on the solution behaviors of the equations.

We now demonstrate that, in general, numerical instabilities always occur in the discrete modeling of ordinary differential equations if one uses either the central difference or the forward Euler schemes provided the non-derivative terms are modeled locally on the computational grid. For our purposes, it is sufficient to prove this for the scalar equation

$$\frac{dy}{dt} = f(y), \qquad (2.7.1)$$

where

$$f(y) = 0, \qquad (2.7.2)$$

is assumed to have only simple zeros. For this autonomous, first-order differential equation, numerical instabilities will occur whenever the linear stability properties of any of the fixed-points for the discrete model differs from those of the differential equation [13, 23, 24, 25].

The fixed-points or constant solutions of Eq. (2.7.1) are solutions to the equation

$$f(\bar{y}) = 0. \qquad (2.7.3)$$

Denote these zeros by $\{\bar{y}^{(i)}\}$, where $i = 1, 2, \ldots, I$. Note that I may be unbounded. Now, define R_i as follows

$$R_i \equiv \frac{df[\bar{y}^{(i)}]}{dy}. \tag{2.7.4}$$

The application of linear stability analysis to the i-th fixed-point gives the following result [26]:

(i) If $R_i > 0$, the fixed-point $y(t) = \bar{y}^{(i)}$ is linearly unstable.

(ii) If $R_i < 0$, the fixed-point $y(t) = \bar{y}^{(i)}$ is linearly stable.

Now construct a central difference discrete model for Eq. (2.7.1), i.e.,

$$\frac{y_{k+1} - y_{k-1}}{2h} = f(y_k). \tag{2.7.5}$$

For small perturbations, ϵ_k, about the fixed-point $\bar{y}^{(i)}$, we have

$$y_k = \bar{y}^{(i)} + \epsilon_k. \tag{2.7.6}$$

If Eq. (2.7.6) is substituted into Eq. (2.7.5) and only linear terms are kept, then we obtain

$$\frac{\epsilon_{k+1} - \epsilon_{k-1}}{2h} = R_i \epsilon_k. \tag{2.7.7}$$

An examination of the characteristic equation for Eq. (2.7.7)

$$r^2 - (2hR_i)r - 1 = 0 \tag{2.7.8}$$

shows that one root is always larger than one in magnitude. In fact,

$$r_\pm = hR_i \pm \sqrt{1 + h^2 R_i^2}. \tag{2.7.9}$$

Since,

$$\epsilon_k = A(r_+)^k + B(r_-)^k, \tag{2.7.10}$$

where A and B are arbitrary, but small constants, we must conclude that the fixed-point at $y_k = \bar{y}^{(i)}$ is linearly unstable. However, if $R_i < 0$, then the corresponding fixed-point of the differential equation is stable. Therefore, the use of the central difference scheme of Eq. (2.7.5) leads to a discrete model of Eq. (2.7.1) for which all the fixed-points are linearly unstable. This means that the central difference scheme has numerical instabilities for all $h > 0$. As stated previously, the main reason for the occurrence of numerical instabilities in this case is that the order of the finite-difference equation is larger than the order of the corresponding differential equation.

Let us now investigate the linear stability properties of the fixed-points for the forward Euler scheme for Eq. (2.7.1). It is given by the following expression

$$\frac{y_{k+1} - y_k}{h} = f(y_k). \tag{2.7.11}$$

A perturbation of the i-th fixed-point, as given by Eq. (2.7.6), leads to the perturbation equation

$$\frac{\epsilon_{k+1} - \epsilon_k}{h} = R_i \epsilon_k, \tag{2.7.12}$$

or

$$\epsilon_{k+1} = (1 + hR_i)\epsilon_k, \tag{2.7.13}$$

which has the solution

$$\epsilon_k = \epsilon_0 (1 + hR_i)^k. \tag{2.7.14}$$

Detailed study of Eq. (2.7.14) gives the following results:

(i) For $R_i > 0$, the fixed-point $\bar{y}^{(i)}$ is linearly unstable for both the differential Eq. (2.7.1) and the difference Eq. (2.7.11) for $h > 0$.

(ii) For $R_i < 0$, which corresponds to a linearly stable fixed-point for the differential Eq. (2.7.1), the fixed-point of the discrete model, namely Eq. (2.7.11), has the properties:

$$0 < h < \frac{2}{|R_i|}, \qquad y_k = \bar{y}^{(i)} \text{ is linearly stable;}$$

$$h \geq \frac{2}{|R_i|}, \qquad y_k = \bar{y}^{(i)} \text{ is linearly unstable.}$$

Consequently, we conclude that the forward Euler scheme and the differential equation will have corresponding fixed-points with the same linear stability properties only if there is a limitation on the step-size h, i.e.,

$$0 < h < h^* = \frac{2}{R^*},\tag{2.7.15}$$

where

$$R^* = \text{Max}\{|R_i|; i = 1, 2, \ldots, I\}.\tag{2.7.16}$$

Numerical instabilities will occur whenever $h > h^*$. This type of numerical instability will be called a threshold instability.

Note that for the central difference scheme $h^* = 0$, i.e., numerical instabilities occur for all $h > 0$.

The previous two finite-difference methods were explicit schemes. We now investigate the properties of an implicit discrete model for Eq. (2.7.1), the backward Euler scheme. It is given by the expression

$$\frac{y_{k+1} - y_k}{h} = f(y_{k+1}).\tag{2.7.17}$$

For small perturbations about the fixed-point at $y_k = \bar{y}^{(i)}$, the equation for ϵ_k is

$$\frac{\epsilon_{k+1} - \epsilon_k}{h} = R_i \epsilon_{k+1},\tag{2.7.18}$$

or

$$\epsilon_{k+1} = \left(\frac{1}{1 - hR_i}\right) \epsilon_k,\tag{2.7.19}$$

which has the solution

$$\epsilon_k = \epsilon_0 \left(\frac{1}{1 - hR_i}\right)^k.\tag{2.7.20}$$

Inspection of Eq. (2.7.20) leads to the following conclusions:

(i) For $R_i < 0$, the fixed-point of Eq. (2.7.17) is linearly stable for all $h > 0$. Thus, the stability properties of the finite-difference scheme and the differential equation are the same.

(ii) For $R_i > 0$, the finite-difference scheme is linearly unstable for

$$0 < h \leq \frac{2}{R_i}, \qquad (2.7.21)$$

but, is linearly stable for

$$h > \frac{2}{R_i}. \qquad (2.7.22)$$

Note that for

$$h > \frac{2}{\bar{R}}, \qquad \bar{R} = \text{Min}\{|R_i|; i = 1, 2, \ldots, I\}, \qquad (2.7.23)$$

all the fixed-points of this implicit scheme are linearly stable. This phenomena is called super-stability by Dahlquist et al. [27] and has been investigated by Lorenz [28], Dieci and Estep [29], and Corless et al. [24]. This phenomena is of great interest since, for systems of ordinary differential equations, there exist discrete models that produce solutions that are not chaotic even though the differential equations themselves are known to have chaotic behavior. This result is the "natural complement of computational chaos" (Corless et al. [24]) or numerical instabilities that can arise when certain finite-difference schemes are used to construct discrete models of ordinary differential equations. Above, we have shown that super-stability can also occur in the backward Euler scheme for a single scalar equation.

The next chapter will be devoted to the study of nonstandard finite-difference schemes and how they can be used to eliminate the elementary forms of numerical instabilities as shown to exist in the present chapter.

References

1. G. Iooss, *Bifurcation of Maps and Applications* (North-Holland, Amsterdam, 1979).

2. V. Arnold, *Geometrical Methods in the Theory of Ordinary Differential Equations* (Springer-Verlag, New York, 1983).

3. G. Iooss and M. Adelmeyer, *Topics in Bifurcation Theory and Applications* (World Scientific, Singapore, 1992).

4. F. B. Hilderbrand, *Finite-Difference Equations and Simulations* (Prentice-Hall; Englewood Cliffs, NJ; 1968).

5. V. D. Barger and M. G. Olsson, *Classical Mechanics: A Modern Perspective* (McGraw-Hill, New York, 1973).

6. R. E. Mickens, *Nonlinear Oscillations* (Cambridge University Press, New York, 1981).

7. R. E. Mickens, *Difference Equations: Theory and Applications* (Van Nostrand Reinhold, New York, 1990).

8. M. K. Jain, *Numerical Solution of Differential Equations* (Wiley, New York, 2nd edition, 1984).

9. J. M. Ortega and W. G. Poole, Jr., *Numerical Methods for Differential Equations* (Pitman; Mashfield, MA; 1981).

10. M. Yamaguti and S. Ushiki, *Physica* **3D**, 618–626 (1981). Chaos in numerical analysis of ordinary differential equations.

11. S. Ushiki, *Physica* **4D**, 407–424 (1982). Central difference scheme and chaos.

12. J. M. Sanz-Serna, *SIAM Journal of Scientific and Statistical Computing* **6**, 923–938 (1985). Studies in numerical nonlinear instability I. Why do leapfrog schemes go unstable?

13. R. E. Mickens, *Dynamic Systems and Applications* **1**, 329–340 (1992). Finite-difference schemes having the correct linear stability properties for all finite step-sizes II.

14. R. M. May, *Nature* **261**, 459–467 (1976). Simple mathematical models with very complicated dynamics.

15. P. Collet and J.-P. Eckmann, *Iterated Maps of the Interval as Dynamical Systems* (Birkhäuser, Boston, 1980).

16. T. Li and J. Yorke, *American Mathematical Monthly* **82**, 985–992 (1975). Period-3 implies chaos.

17. R. L. Devaney, *An Introduction to Chaotic Dynamical Systems* (Benjamin/Cummings; Menlo Park, CA; 1986).

18. G. B. Whitham, *Linear and Nonlinear Waves* (Wiley-Interscience, New York, 1974).

19. J. M. Burgers, *Advanced in Applied Mechanics* 1, 171–199 (1948). A mathematical model illustrating the theory of turbulence.

20. R. E. Mickens, Difference equation models of differential equations having zero local truncation errors, in *Differential Equations*, I. W. Knowles and R. T. Lewis, editors (North-Holland, Amsterdam, 1984), pp. 445–449.

21. R. E. Mickens, Mathematical modeling of differential equations by difference equations, in *Computational Acoustics: Wave Propagation*, D. Lee et al., editors (Elsevier Science Publications B.V., Amsterdam, 1988), pp. 387–393.

22. R. E. Mickens, *Numerical Methods for Partial Differential Equations* 5, 313–325 (1989). Exact solutions to a finite-difference model for a nonlinear reaction-advection equation: Implications for numerical analysis.

23. R. E. Mickens, Runge-Kutta schemes and numerical instabilities: The Logistic equation, in *Differential Equations and Mathematical Physics*, I. Knowles and Y. Saito, editors (Springer-Verlag, Berlin, 1987), pp. 337–341.

24. R. M. Corless, C. Essex and M. A. H. Nerenberg, *Physics Letters* **A157**, 27–36 (1991). Numerical methods can suppress chaos.

25. A. Iserles, A. T. Peplow and A. M. Stuart, *SIAM Journal of Numerical Analysis* **28**, 1723–1751 (1991). A unified approach to spurious solutions introduced by time discretization. Part I: Basic theory.

26. M. Sever, *Ordinary Differential Equations* (Boole Press, Dublin, 1987), pp. 101–103.

27. G. Dahlquist, L. Edsberg, G. Skollermo, and G. Soderlind, Are the numerical methods and software satisfactory for chemical kinetics, in *Numerical Integration of Differential Equations and Large Linear Systems*, J. Hinze, editor (Springer-Verlag, Berlin, 1982), pp. 149–164.

28. E. N. Lorentz, *Physica* **D35**, 299–317 (1989). Computational chaos — A prelude to computational instability.

29. L. Dieci and D. Estep, Georgia Institute of Technology, Tech. Rep. Math. 050290-039 (1990). Some stability aspects of schemes for the adaptive integration of stiff initial value problems.

Chapter 3
NONSTANDARD FINITE-DIFFERENCE SCHEMES

3.1 Introduction

This chapter provides background information to understand the general rules of Mickens [1] for the construction of nonstandard finite-difference schemes for differential equations. First, the concept of an exact difference scheme is introduced and defined. Second, a theorem is stated and proved that all ordinary differential equations have a unique exact difference scheme. The major consequence of this result is that such schemes do not allow numerical instabilities to occur. Third, using this theorem, exact difference schemes are constructed for a variety of both ordinary and partial differential equations. From these results are formulated a set of modeling rules for the construction of nonstandard finite-difference schemes. Fourth, the notion of best difference schemes is defined and its use in the actual construction of finite-difference schemes is illustrated by several examples.

Before proceeding, we would like to make several comments related to the discrete modeling of the scalar ordinary differential equation

$$\frac{dy}{dt} = f(y, \lambda), \tag{3.1.1}$$

where λ is an n-parameter vector. The most general finite-difference model for Eq. (3.1.1) that is of first-order in the discrete derivative takes the following form

$$\frac{y_{k+1} - y_k}{\phi(h, \lambda)} = F(y_k, y_{k+1}, \lambda, h). \tag{3.1.2}$$

The discrete derivative, on the left-side, is a generalization [2] of that which is normally used, namely [3],

$$\frac{dy}{dt} \rightarrow \frac{y_{k+1} - y_k}{h}. \tag{3.1.3}$$

From Eq. (3.1.2), we have

$$\frac{dy}{dt} \rightarrow \frac{y_{k+1} - y_k}{\phi(h, \lambda)},$$ (3.1.4)

where the *denominator function* $\phi(h, \lambda)$ has the property [2]

$$\phi(h, \lambda) = h + O(h^2),$$

$$\lambda = \text{fixed}, \qquad h \rightarrow 0.$$ (3.1.5)

This form for the discrete derivative is based on the traditional definition of the derivative which can be generalized as follows:

$$\frac{dy}{dt} = \lim_{h \rightarrow 0} \frac{y[t + \psi_1(h)] - y(t)}{\psi_2(h)},$$ (3.1.6)

where

$$\psi_i(h) = h + O(h^2), \qquad h \rightarrow 0; \quad i = 1, 2.$$ (3.1.7)

Examples of functions $\psi(h)$ that satisfy this condition are

$$\psi(h) = \begin{cases} h, \\ \sin(h), \\ e^h - 1, \\ 1 - e^{-h}. \\ \frac{1 - e^{-\lambda h}}{\lambda}, \\ \text{etc.} \end{cases}$$

Note that in taking the $\lim h \rightarrow 0$ to obtain the derivative, the use of any of these $\psi(h)$ will lead to the usual result for the first derivative

$$\frac{dy}{dt} = \lim_{h \rightarrow 0} \frac{y[t + \psi_1(h)] - y(t)}{\psi_2(h)} = \lim_{h \rightarrow 0} \frac{y(t + h) - y(t)}{h}.$$ (3.1.8)

However, for h finite, these discrete derivatives will differ greatly from those conventionally given in the literature, such as Eq. (3.1.3). This fact not only allows for the construction of a larger class of finite-difference models, but also provides for more ambiguity in the modeling process.

3.2 Exact Finite-Difference Schemes

We consider only first-order, scalar ordinary differential equations in this section. However, the results can be easily generalized to coupled systems of first-order ordinary differential equations.

It should be acknowledged that the early work of Potts [4] played a fundamental role in interesting the author in the concept of exact finite-difference schemes.

Consider the general first-order differential equation

$$\frac{dy}{dt} = f(y, t, \lambda), \qquad y(t_0) = y_0, \tag{3.2.1}$$

where $f(y, t, \lambda)$ is such that Eq. (3.2.1) has a unique solution over the interval, $0 \le t < T$ [5, 6] and for λ in the interval $\lambda_1 \le \lambda \le \lambda_2$. (For dynamical systems of interest, in general, $T = \infty$, i.e., the solution exists for all time.) This solution can be written as

$$y(t) = \phi(\lambda, y_0, t_0, t), \tag{3.2.2}$$

with

$$\phi(\lambda, y_0, t_0, t_0) = y_0. \tag{3.2.3}$$

Now consider a discrete model of Eq. (3.2.1)

$$y_{k+1} = g(\lambda, h, y_k, t_k), \qquad t_k = hk. \tag{3.2.4}$$

Its solution can be expressed in the form

$$y_k = \psi(\lambda, h, y_0, t_0, t_k), \tag{3.2.5}$$

with

$$\psi(\lambda, h, y_0, t_0, t_0) = y_0. \tag{3.2.6}$$

Definition 1. Equations (3.2.1) and (3.2.4) are said to have the *same general solution* if and only if

$$y_k = y(t_k) \tag{3.2.7}$$

for arbitrary values of h.

Definition 2. An *exact difference scheme* is one for which the solution to the difference equation has the same general solution as the associated differential equation.

These definitions lead to the following result:

Theorem. *The differential equation*

$$\frac{dy}{dt} = f(y, t, \lambda), \qquad y(t_0) = y_0, \tag{3.2.8}$$

has an exact finite-difference scheme given by the expression

$$y_{k+1} = \phi[\lambda, y_k, t_k, t_{k+1}], \tag{3.2.9}$$

where ϕ is that of Eq. (3.2.2).

Proof [7]. The group property [5, 6] of the solutions to Eq. (3.2.8) gives

$$y(t + h) = \phi[\lambda, y(t), t, t + h]. \tag{3.2.10}$$

If we now make the identifications

$$t \to t_k, \qquad y(t) \to y_k, \tag{3.2.11}$$

then Eq. (3.2.10) becomes

$$y_{k+1} = \phi(\lambda, y_k, t_k, t_{k+1}). \tag{3.2.12}$$

This is the required ordinary difference equation that has the same general solution as Eq. (3.2.8).

Comments. (i) If all solutions of Eq. (3.2.8) exist for all time, i.e., $T = \infty$, then Eq. (3.2.10) holds for all t and h. Otherwise, the relation is assumed to hold whenever the right-side of Eq. (3.2.10) is well defined.

(ii) The theorem is only an existence theorem. It basically says that if an ordinary differential equation has a solution, then an exact finite-difference scheme exists. In general, no guidance is given as to how to actually construct such a scheme.

(iii) A major implication of the theorem is that the solution of the difference equation is exactly equal to the solution of the ordinary differential equation on the computational grid for fixed, but, arbitrary step-size h.

(iv) The theorem can be easily generalized to systems of coupled, first-order ordinary differential equations.

The question now arises as to whether exact difference schemes exist for partial differential equations. For an arbitrary partial differential equation the answer is (probably) no. This negative result is a consequence of the fact that given an arbitrary partial differential equation there exists no clear, unambiguous accepted definition of a general solution to the equation [8, 9]. However, we should expect that certain classes of partial differential equations will have exact difference models. Note that in this case some type of functional relation should exist between the various (space and time) step-sizes.

The discovery of exact discrete models for particular ordinary and partial differential equations is of great importance, primarily because it allows us to gain insights into the better construction of finite-difference schemes. They also provide the computational investigator with useful benchmarks for comparison with the standard procedures.

3.3 Examples of Exact Schemes

In this section, we will use the theorem of the last section "in reverse" to construct exact difference schemes for several ordinary and partial differential equations for which exact general solutions are explicitly known. These schemes have the property that their solutions do not have numerical instabilities.

However, before proceeding, it should be indicated that given a set of linearly independent functions

$$\{y^{(i)}(t)\}; \qquad i = 1, 2, \ldots, N, \tag{3.3.1}$$

it is always possible to construct an N-th order linear difference equation that has the corresponding discrete functions as solutions [10]. For let

$$y_k^{(i)} \equiv y^{(i)}(t_k), \qquad t_k = (\Delta t)k = hk; \tag{3.3.2}$$

then the following determinant gives the required difference equation

$$\begin{vmatrix} y_k & y_k^{(1)} & y_k^{(2)} & \cdots & y_k^{(n)} \\ y_{k+1} & y_{k+1}^{(1)} & y_{k+1}^{(2)} & \cdots & y_{k+1}^{(n)} \\ \vdots & \vdots & \vdots & & \vdots \\ y_{k+n} & y_{k+n}^{(1)} & y_{k+n}^{(2)} & \cdots & y_{k+n}^{(n)} \end{vmatrix} = 0. \tag{3.3.3}$$

As a first example to illustrate this procedure, consider the single function

$$y^{(1)}(t) = e^{-\lambda t}. \tag{3.3.4}$$

This is (with an arbitrary multiplicative constant) the general solution to the first-order differential equation

$$\frac{dy}{dt} = -\lambda y. \tag{3.3.5}$$

The corresponding difference equation is

$$\begin{vmatrix} y_k & y_k^{(1)} \\ y_{k+1} & y_{k+1}^{(1)} \end{vmatrix} = \begin{vmatrix} y_k & e^{-\lambda hk} \\ y_{k+1} & e^{-\lambda h(k+1)} \end{vmatrix} = e^{-\lambda hk} \begin{vmatrix} y_k & 1 \\ y_{k+1} & e^{-\lambda h} \end{vmatrix}$$
$$= e^{-\lambda hk} \left[e^{-\lambda h} y_k - y_{k+1} \right] = 0, \tag{3.3.6}$$

or

$$y_{k+1} = e^{-\lambda h} y_k. \tag{3.3.7}$$

This is the exact difference equation corresponding to Eq. (3.3.5). However, a more instructive form can be obtained by carrying out the following manipulations:

$$y_{k+1} - y_k = (e^{-\lambda h} - 1)y_k = -\lambda \left(\frac{1 - e^{-\lambda h}}{\lambda} \right) y_k, \tag{3.3.8}$$

and finally,

$$\frac{y_{k+1} - y_k}{\left(\frac{1 - e^{-\lambda h}}{\lambda}\right)} = -\lambda y_k. \tag{3.3.9}$$

Note that the standard forward Euler scheme for this differential equation is

$$\frac{y_{k+1} - y_k}{h} = -\lambda y_k. \tag{3.3.10}$$

For a second example, consider the harmonic oscillator differential equation

$$\frac{d^2 y}{dt^2} + \omega^2 y = 0, \tag{3.3.11}$$

where ω is a real constant. The two linearly independent solutions are

$$y^{(1)}(t) = \cos(\omega t), \qquad y^{(2)}(t) = \sin(\omega t), \tag{3.3.12}$$

or in complex form

$$\bar{y}^{(1)}(t) = e^{i\omega t}, \qquad \bar{y}^{(2)}(t) = e^{-i\omega t}. \tag{3.3.13}$$

Therefore,

$$\begin{vmatrix} y_k & e^{i\omega h k} & e^{-i\omega h k} \\ y_{k+1} & e^{i\omega h(k+1)} & e^{-i\omega h(k+1)} \\ y_{k+2} & e^{i\omega h(k+2)} & e^{-i\omega h(k+2)} \end{vmatrix} = 0, \tag{3.3.14}$$

and

$$y_{k+2} - [2\cos(\omega h)]y_{k+1} + y_k = 0. \tag{3.3.15}$$

Shifting downward the index k by one unit and using the identity

$$2\cos(\omega h) = 2 - 4\sin^2\left(\frac{\omega h}{2}\right), \tag{3.3.16}$$

Eq. (3.3.15) can be put in the form

$$\frac{y_{k+1} - 2y_k + y_{k-1}}{\left(\frac{4}{\omega^2}\right)\sin^2\left(\frac{h\omega}{2}\right)} + \omega^2 y_k = 0. \tag{3.3.17}$$

This is the exact finite-difference scheme for Eq. (3.3.11) and should be compared to the standard central difference model of the harmonic oscillator differential equation

$$\frac{y_{k+1} - 2y_k + y_{k-1}}{h^2} + \omega^2 y_k = 0. \tag{3.3.18}$$

For nonlinear differential equations, the above procedure cannot be used to construct exact finite-difference schemes. A procedure based on the theorem of the previous section must be used. The following outlines the steps to be applied:

(i) Consider a system of N coupled, first-order, ordinary differential equations

$$\frac{dY}{dt} = F(Y, t, \lambda), \qquad Y(t_0) = Y_0, \tag{3.3.19}$$

where Y, F are N-dimensional column vectors whose i-th components are

$$(Y)_i = y^{(i)}(t), \tag{3.3.20}$$

$$(F)_i = f^{(i)}[y^{(1)}, y^{(2)}, \dots, y^{(N)}; t, \lambda]. \tag{3.3.21}$$

(ii) Denote the general solution to Eq. (3.3.19) by

$$Y(t) = \Phi(\lambda, Y_0, t_0, t) \tag{3.3.22}$$

where

$$y^{(i)}(t) = \phi^{(i)}[\lambda, y_0^{(1)}, y_0^{(2)}, \dots, y_0^{(N)}, t_0, t]. \tag{3.3.23}$$

(iii) The exact difference equation corresponding to the differential equation is obtained by making the following substitutions in Eq. (3.3.22):

$$\begin{cases} Y(t) \to Y_{k+1}, \\ Y_0 = Y(t_0) \to Y_k, \\ t_0 \to t_k, \\ t \to t_{k+1}. \end{cases} \tag{3.3.24}$$

As a first illustration of the procedure, consider again the decay differential equation of Eq. (3.3.5). The general solution is

$$y(t) = y_0 e^{-\lambda(t-t_0)}. \tag{3.3.25}$$

The substitutions of Eq. (3.3.24) give

$$y_{k+1} = y_k e^{-\lambda h} \tag{3.3.26}$$

which is just Eq. (3.3.7).

For our second example, consider the general Logistic differential equation

$$\frac{dy}{dt} = \lambda_1 y - \lambda_2 y^2, \qquad y(t_0) = y_0, \tag{3.3.27}$$

where λ_1 and λ_2 are constants. The solution to the initial value problem of Eq. (3.3.27) is given by the following expression

$$y(t) = \frac{\lambda_1 y_0}{(\lambda_1 - y_0 \lambda_2) e^{-\lambda_1 (t - t_0)} + \lambda_2 y_0}. \tag{3.3.28}$$

Making the substitutions of Eq. (3.3.24) gives

$$y_{k+1} = \frac{\lambda_1 y_k}{(\lambda_1 - \lambda_2 y_k) e^{-\lambda_1 h} + \lambda_2 y_k}. \tag{3.3.29}$$

Additional algebraic manipulation gives

$$\frac{y_{k+1} - y_k}{\left(\frac{e^{\lambda_1 h} - 1}{\lambda_1}\right)} = \lambda_1 y_k - \lambda_2 y_{k+1} y_k. \tag{3.3.30}$$

Again, note that this form does not correspond to any of the discrete models constructed in the previous chapter using standard methods.

Observe, with $\lambda_2 = 0$ and $\lambda_1 \to -\lambda$, that Eq. (3.3.30) goes to the relation of Eq. (3.3.9). Also, we can obtain the exact difference scheme for the differential equation

$$\frac{dy}{dt} = -y^2 \tag{3.3.31}$$

by setting, in Eq. (3.3.30), $\lambda_1 = 0$ and $\lambda_2 = 1$. This gives the exact difference scheme

$$\frac{y_{k+1} - y_k}{h} = -y_{k+1} y_k. \tag{3.3.32}$$

The harmonic oscillator equation

$$\frac{d^2y}{dt^2} + y = 0,$$

(3.3.33)

can be written as a system of two coupled, first-order differential equations

$$\frac{dy^{(1)}}{dt} = y^{(2)},$$

(3.3.34a)

$$\frac{dy^{(2)}}{dt} = -y^{(1)},$$

(3.3.34b)

where $y^{(1)}(t) = y(t)$. With the initial conditions

$$y_0^{(1)} = y^{(1)}(t_0), \qquad y_0^{(2)} = y^{(2)}(t_0).$$

(3.3.35)

Equations (3.3.34) have the solutions

$$y^{(1)}(t) = \left(\frac{1}{2}\right)\left[y_0^{(1)} - iy_0^{(2)}\right]e^{i(t-t_0)} + \left(\frac{1}{2}\right)\left[y_0^{(1)} + iy_0^{(2)}\right]e^{-i(t-t_0)},$$

(3.3.36)

$$iy^{(2)}(t) = -\left(\frac{1}{2}\right)\left[y_0^{(1)} - iy_0^{(2)}\right]e^{i(t-t_0)} + \left(\frac{1}{2}\right)\left[y_0^{(1)} + iy_0^{(2)}\right]e^{-i(t-t_0)}.$$

(3.3.37)

Making the substitutions of Eq. (3.3.24) gives

$$y_{k+1}^{(1)} = \cos(h)y_k^{(1)} + \sin(h)y_k^{(2)},$$

(3.3.38)

$$y_{k+1}^{(2)} = \sin(h)y_k^{(1)} + \cos(h)y_k^{(2)}.$$

(3.3.39)

Finally, eliminating $y_k^{(2)}$ gives the expression

$$\frac{y_{k+1} - 2y_k + y_{k-1}}{4\sin^2\left(\frac{h}{2}\right)} + y_k = 0,$$

(3.3.40)

which is the exact finite-difference scheme for the harmonic oscillator. Note that if $h \to \omega h$, then Eq. (3.3.40) becomes Eq. (3.3.17).

Without giving the details, we now present several other ordinary differential equations and their exact discrete models [11]:

$$2\frac{dy}{dt} + y = \frac{1}{y},$$

(3.3.41a)

$$2\left[\frac{y_{k+1} - y_k}{1 - e^{-h}}\right] + \frac{y_k^2}{\left(\frac{y_{k+1}+y_k}{2}\right)} = \frac{1}{\left(\frac{y_{k+1}+y_k}{2}\right)}; \tag{3.3.41b}$$

$$\frac{dy}{dt} = -y^3, \tag{3.3.42a}$$

$$\frac{y_{k+1} - y_k}{h} = -\left[\frac{2y_{k+1}}{y_{k+1} + y_k}\right] y_{k+1} y_k^2; \tag{3.3.42b}$$

$$\frac{d^2 y}{dt^2} = \lambda \frac{dy}{dt}, \tag{3.3.43a}$$

$$\frac{y_{k+1} - 2y_k + y_{k-1}}{\left(\frac{e^{\lambda h}-1}{\lambda}\right)h} = \lambda \left(\frac{y_k - y_{k-1}}{h}\right). \tag{3.3.43b}$$

All of the above examples of exact finite-difference schemes have been obtained for ordinary differential equations. We now turn to an example of a partial differential equation for which an exact discrete model exists.

Consider the nonlinear reaction-advection equation

$$u_t + u_x = u(1 - u), \tag{3.3.44}$$

with the initial value

$$u(x, 0) = f(x), \tag{3.3.45}$$

where $f(z)$ is bounded with a bounded derivative. The nonlinear transformation [1]

$$u(x, t) = \frac{1}{w(x, t)} \tag{3.3.46}$$

reduces Eq. (3.3.44) to the linear equation

$$w_t + w_x = 1 - w. \tag{3.3.47}$$

The general solution of this equation can be easily determined by standard methods [8]. It is

$$w(x, t) = g(x - t)e^{-t} + 1, \tag{3.3.48}$$

where $g(z)$ is an arbitrary function of z having a bounded first derivative. Imposing the initial condition of Eq. (3.3.45) allows g to be calculated, i.e.,

$$g(x) + 1 = \frac{1}{f(x)} \tag{3.3.49}$$

or

$$g(x) = \frac{1 - f(x)}{f(x)}. \tag{3.3.50}$$

Using this result with Eqs. (3.3.46) and (3.3.48), we can obtain the solution to Eqs. (3.3.44) and (3.3.45); it is given by the expression

$$u(x, t) = \frac{f(x - t)}{e^{-t} + (1 - e^{-t})f(x - t)}. \tag{3.3.51}$$

To proceed, we first construct the exact finite-difference scheme for the unidirectional wave equation

$$u_t + u_x = 0. \tag{3.3.52}$$

The general solution of this equation is [8]

$$u(x, t) = H(x - t), \tag{3.3.53}$$

where H is an arbitrary function. Now the partial difference equation

$$u_m^{k+1} = u_{m-1}^k \tag{3.3.54}$$

has as its general solution an arbitrary function of $(m - k)$ [10], i.e.,

$$u_m^k = F(m - k). \tag{3.3.55}$$

If we impose the condition

$$\Delta x = \Delta t, \tag{3.3.56}$$

then Eq. (3.3.54) can be rewritten in the following form

$$\frac{u_m^{k+1} - u_m^k}{\beta(\Delta t)} + \frac{u_m^k - u_{m-1}^k}{\beta(\Delta x)} = 0, \tag{3.3.57}$$

where $\beta(z)$ has the property

$$\beta(z) = z + O(z^2), \qquad z \to 0. \tag{3.3.58}$$

The general solution of Eq. (3.3.57), which is formally equivalent to Eq. (3.3.54), is

$$u_m^k = F_1[h(m - k)] \qquad (h = \Delta x = \Delta t)$$

$$= F_1(x_m - t_k), \tag{3.3.59}$$

where F_1 is an arbitrary function of its argument. Thus, the exact finite-difference scheme for the unidirectional wave equation is Eq. (3.3.57).

We can use this result to calculate the exact difference scheme for Eq. (3.3.44). Solving Eq. (3.3.51) for $f(x - t)$ gives

$$f(x - t) = \frac{e^{-t}u(x,t)}{1 - (1 - e^{-t})u(x,t)}. \tag{3.3.60}$$

Now make the following substitutions in the last equation

$$\begin{cases} x \to x_m = (\Delta x)m, \qquad t \to t_k = (\Delta t)k, \qquad \Delta x = \Delta t = h, \\ u(x,t) \to u_m^k, \\ f(x - t) \to f[h(m - k)] = f_m^k. \end{cases} \tag{3.3.61}$$

Doing this gives

$$f_m^k = \frac{e^{-hk}u_m^k}{1 - (1 - e^{-hk})u_m^k}. \tag{3.3.62}$$

However, from Eqs. (3.3.54) and (3.3.55), we know that f_m^k satisfies the following partial difference equation

$$f_m^{k+1} = f_{m-1}^k. \tag{3.3.63}$$

Therefore, we have

$$\frac{e^{-k(k+1)}u_m^{k+1}}{1 - [1 - e^{-k(k+1)}]u_m^{k+1}} = \frac{e^{-kk}u_{m-1}^k}{1 - (1 - e^{-hk})u_{m-1}^k}. \tag{3.3.64}$$

After some algebraic manipulations, this expression becomes

$$\frac{u_m^{k+1} - u_m^k}{e^{\Delta t} - 1} + \frac{u_m^k - u_{m-1}^k}{e^{\Delta x} - 1} = u_{m-1}^k(1 - u_m^{k+1}), \tag{3.3.65a}$$

$$\Delta t = \Delta x. \qquad (3.3.65b)$$

Discrete models of the nonlinear reaction-advection equation using the standard rules do not have the structure of Eqs. (3.3.65). For example, a particular standard model is

$$\frac{u_m^{k+1} - u_m^k}{\Delta t} + \frac{u_{m+1}^k - u_m^k}{\Delta x} = u_m^k(1 - u_m^k). \qquad (3.3.66)$$

3.4 Nonstandard Modeling Rules

Let us now examine in detail the results obtained in the previous section. In particular, we concentrate on the exact finite-difference schemes for the general Logistic ordinary differential equation and the nonlinear, reaction-advection partial differential equation. These are given, respectively, by Eqs. (3.3.27) and (3.3.30), and (3.3.44) and (3.3.65).

The following observations are important:

(i) Exact finite-difference schemes generally require that nonlinear terms be modeled nonlocally. Thus, for the Logistic equation the y^2 term is evaluated at two different grid points

$$y^2 \rightarrow y_{k+1}y_k. \qquad (3.4.1)$$

Similarly, the u^2 term for the nonlinear, reaction-advection equation is modeled by the expression

$$u^2 \rightarrow u_{m-1}^k u_m^{k+1}. \qquad (3.4.2)$$

This corresponds to u^2 being evaluated at two different lattice space-points and two different lattice time-points. Note that

$$\operatorname*{Lim}_{\substack{h \to 0 \\ k \to \infty \\ hk=t=\text{fixed}}} y_{k+1}y_k = \operatorname*{Lim}_{\substack{h \to 0 \\ k \to \infty \\ hk=t=\text{fixed}}} y_k^2 = y(t), \qquad (3.4.3)$$

and

$$\underset{\substack{\Delta x \to 0 \\ \Delta t \to 0 \\ k \to \infty \\ m \to \infty \\ (\Delta x)m=x=\text{fixed} \\ (\Delta t)k=t=\text{fixed}}}{\text{Lim}} u_{m-1}^k u_m^{k+1} = \underset{\substack{\Delta x \to 0 \\ \Delta t \to 0 \\ k \to \infty \\ m \to \infty \\ (\Delta x)m=x=\text{fixed} \\ (\Delta t)k=t=\text{fixed}}}{\text{Lim}} (u_m^k)^2 = [u(x,t)]^2. \qquad (3.4.4)$$

However, for finite, fixed, nonzero values of the step-sizes, the two representations of the squared terms in Eqs. (3.4.3) and (3.4.4) are not equal, i.e.,

$$y_{k+1} y_k \neq (y_k)^2, \qquad (3.4.5a)$$

$$u_{m-1}^k u_m^{k+1} \neq (u_m^k)^2. \qquad (3.4.5b)$$

Therefore, a seemingly trivial modification in the modeling of nonlinear terms can lead to major changes in the solution behaviors of the difference equations.

(ii) The discrete derivatives for both differential equations have denominator functions that are more complicated than those used in the standard modeling procedure. For example, the time-derivative in the Logistic equation is replaced by the following discrete representation

$$\frac{dy}{dt} \to \frac{y_{k+1} - y_k}{\left(\frac{e^{\lambda_1 h} - 1}{\lambda_1} \right)}. \qquad (3.4.6)$$

Thus, the denominator function depends on both the parameter λ_1 and the step-size $h = \Delta t$.

(iii) In the discrete modeling of partial differential equations, functional relations may exist between the various step-sizes. For the nonlinear, reaction-advection equation, the required restriction is $\Delta t = \Delta x$.

(iv) Of importance is the observation that for partial differential equations, the modeling of first-derivatives may require the use of a forward Euler type discrete derivative for the time variable, but, a backward Euler type discrete derivative for the space variable. See Eq. (3.3.65).

(v) Finally, we found that the order of the discrete derivatives in exact finite-difference schemes is always equal to the corresponding order of the derivatives of the differential equation.

With the above facts in hand, we now study the various sources of numerical instabilities for standard models of the Logistic and nonlinear, reaction-advection equations. First, consider the following finite-difference scheme for the Logistic equation

$$\frac{y_{k+1} - y_k}{h} = y_k(1 - y_k). \tag{3.4.7}$$

This discrete representation is expected to have numerical instabilities for two reasons: (a) the denominator function is incorrect; (b) the nonlinear term is modeled locally on the grid. See Eq. (3.3.30) for comparison with the exact scheme. Now consider the following model for the nonlinear, reaction-advection equation

$$\frac{u_m^{k+1} - u_m^k}{\Delta t} + \frac{u_{m+1}^k - u_{m-1}^k}{2\Delta x} = u_m^k(1 - u_m^k). \tag{3.4.8}$$

There are several sources of numerical instabilities: (a) The nonlinear term is modeled locally on the computational grid. (b) The first-order space derivative is modeled by a higher order central difference scheme. (c) There is no explicit relation indicated between the space and time step-sizes. Again, comparison to Eq. (3.3.65) should be made.

These results can be used to understand the findings of Mitchell and Bruch [12] who consider the one-dimensional, nonlinear, reaction-diffusion equation

$$u_t = Du_{xx} + \alpha u(1 - u), \tag{3.4.9}$$

where D and α are non-negative constants. This equation is known as the Fisher equation [13]. In our notation, they numerically investigated the properties of the solutions to the finite-difference scheme

$$\frac{u_m^{k+1} - u_m^k}{\Delta t} = D\left[\frac{u_{m+1}^k - 2u_m^k + u_{m-1}^k}{(\Delta x)^2}\right] + \alpha u_m^k(1 - u_m^k). \tag{3.4.10}$$

They found numerical solutions that were chaotic as well as other solutions that diverged. From the perspective of our analysis, it should be clear that these numerical instabilities were a primary consequence of the local modeling for the u^2 term. Note that in Eq. (3.4.10), the discrete space independent difference equation is the Logistic difference equation.

Based on both analytical and numerical studies of exact finite-difference schemes for a large number of ordinary and partial differential equations [1, 2, 3, 11], we present the following rules for the construction of discrete models.

Rule 1. The orders of the discrete derivatives must be exactly equal to the orders of the corresponding derivatives of the differential equations.

Rule 2. Denominator functions for the discrete derivatives must, in general, be expressed in terms of more complicated functions of the step-sizes than those conventionally used.

Rule 3. Nonlinear terms must, in general, be modeled nonlocally on the computational grid or lattice.

Rule 4. Special solutions of the differential equations should also be special (discrete) solutions of the finite-difference models.

Rule 5. The finite-difference equations should not have solutions that do not correspond exactly to solutions of the differential equations.

A major advantage of having an exact difference equation model for a differential equation is that questions related to the usual considerations of consistency, stability and convergence [9, 14, 15, 16] need not arise. However, it is essentially impossible to construct an exact discrete model for an arbitrary differential equation. This is because to do so would be tantamount to knowing the general solution of the original differential equation. However, the situation is not hopeless. The above five modeling rules can be applied to the construction of finite-difference schemes. While these discrete models, in general, will not be exact schemes, they will possess certain very desirable properties. In particular, we may hope to eliminate a number of the problems related to numerical instabilities.

The next section introduces the notion of a *best finite difference scheme*. After discussion of this concept, we present the construction of best discrete models for two nonlinear differential equations.

3.5 Best Finite-Difference Schemes

A *best finite-difference scheme* is a discrete model of a differential equation that is constructed according to the five rules given in Section 3.4. In general, best schemes are not exact schemes. However, they offer the prospect of obtaining finite-difference models that do not possess the standard numerical instabilities. As will be demonstrated in two examples, the application of the five nonstandard modeling rules does not necessarily lead to a unique discrete model for a given differential equation. We are currently studying how to resolve this difficulty.

An equation of fundamental importance in the study of one dimensional, non-linear oscillatory phenomena is the Duffing equation [17]

$$\frac{d^2y}{dt^2} + a\frac{dy}{dt} + by + cy^3 = F\cos\omega t, \tag{3.5.1}$$

where (a, b, c, F, ω) are constants. For our purposes, the following special case will be studied [18]

$$\frac{d^2y}{dt^2} + \omega^2 y + \lambda y^3 = 0, \tag{3.5.2}$$

where ω is the angular frequency of the linear oscillation and λ is a measure of the strength of the nonlinear term. A first-integral or energy relation is

$$\left(\frac{1}{2}\right)\left(\frac{dy}{dt}\right)^2 + \frac{\omega^2 y^2}{2} + \frac{\lambda y^4}{4} = E, \tag{3.5.3}$$

where E is the energy constant. If λ is restricted to be non-negative, i.e.,

$$\lambda \geq 0, \tag{3.5.4}$$

then it follows from Eq. (3.5.3) that all the solutions of Eq. (3.5.2) are bounded and periodic [17].

A standard finite-difference model for Eq. (3.5.2) is

$$\frac{y_{k+1} - 2y_k + y_{k-1}}{h^2} + \omega^2 y_k + \lambda y_k^3 = 0, \qquad (3.5.5)$$

where $h = \Delta t$. This equation can be rewritten as

$$\frac{(y_{k+1} - y_k) - (y_k - y_{k-1})}{h^2} = -\omega^2 v_k - \lambda y_k^3. \qquad (3.5.6)$$

Multiplying by

$$(y_{k+1} - y_k) + (y_k - y_{k-1}) = y_{k+1} - y_{k-1} \qquad (3.5.7)$$

gives

$$\frac{(y_{k+1} - y_k)^2}{h^2} - \frac{(y_k - y_{k-1})^2}{h^2} = -\omega^2 (y_{k+1}y_k - y_k y_{k-1})$$
$$- \lambda(y_{k+1}y_k^3 - y_{k-1}y_k^3). \qquad (3.5.8)$$

The transposition of certain terms gives the following expression

$$\frac{(y_{k+1} - y_k)^2}{h^2} + \omega^2 y_{k+1}y_k + \lambda y_{k+1}y_k^3$$
$$= \frac{(y_k - y_{k-1})^2}{h^2} + \omega^2 y_k y_{k-1} + \lambda y_{k-1}y_k^3. \qquad (3.5.9)$$

If energy is to be conserved, then the right-side of this equation should reduce to the terms on the left-side when k is replaced by $k + 1$. The first two terms do have this property; however, the third term on the right-side does not become the third term on the left-side under this transformation. Therefore, we conclude that the standard finite-difference scheme of Eq. (3.5.5) does not conserve energy [18].

The application of the five rules from the previous section gives the following discrete model for Eq. (3.5.2):

$$\frac{y_{k+1} - 2y_k + y_{k-1}}{\psi} + \omega^2 y_k + \lambda y_k^2 \left(\frac{y_{k+1} + y_{k-1}}{2} \right) = 0, \qquad (3.5.10)$$

where ψ has the property that

$$\psi(h,\omega,\lambda) = h^2 + O(h^4), \qquad h \to 0. \qquad (3.5.11)$$

Note that in the limits

$$h \to 0, \qquad k \to \infty, \qquad hk = t = \text{constant}, \qquad (3.5.12)$$

this finite-difference equation converges to the original Duffing equation as given by Eq. (3.5.2). Also, observe that the nonlinear term y^3 in the differential equation has the following discrete representation

$$y^3 \to y_k^2 \left(\frac{y_{k+1} + y_{k-1}}{2} \right). \qquad (3.5.13)$$

To proceed further, it must now be shown that the discrete model of Eq. (3.5.10) satisfies a conservation law. This is easily done by following the same steps as presented above for Eq. (3.5.5). We find the result

$$\frac{(y_{k+1} - y_k)^2}{2\psi} + \left(\frac{1}{2}\right)\omega^2 y_{k+1}y_k + \left(\frac{1}{4}\right)\lambda y_{k+1}^2 y_k^2$$
$$= \frac{(y_k - y_{k-1})^2}{2\psi} + \left(\frac{1}{2}\right)\omega^2 y_k y_{k-1} + \left(\frac{1}{4}\right)\lambda y_k^2 y_{k-1}^2. \qquad (3.5.14)$$

Observe that the transformation $k \to k+1$ changes the right-side of Eq. (3.5.14) into the expression on the left-side. This means that, independent of the value of k, each side of Eq. (3.5.14) is equal to the same constant. Consequently, the discrete model of the Duffing equation, given by Eq. (3.5.10), has the following associated conservation law

$$\left(\frac{1}{2}\right)\frac{(y_{k+1} - y_k)^2}{\psi} + \left(\frac{1}{2}\right)\omega^2 y_{k+1}y_k + \left(\frac{\lambda}{4}\right)y_{k+1}^2 y_k^2 = \text{constant}. \qquad (3.5.15)$$

This is to be compared to the energy relation of the differential equation as expressed by Eq. (3.5.3).

An ambiguity in the above modeling process is that the denominator function ψ is not uniquely specified. At this level of analysis, any function that obeys the relation given by Eq. (3.5.11) works. Finally, it should be indicated that Potts [19] has also investigated various nonstandard finite-difference approximations to the unforced, undamped Duffing differential equation.

For our second example of the construction of a best finite-difference scheme, we consider the nonlinear diffusion equation [11]

$$u_t = u u_{xx}. \tag{3.5.16}$$

No known exact solution exists for the general initial-value problem for this equation. However, a special rational solution is known. To obtain it, write $u(x,t)$ in the form

$$u(x,t) = C(t)D(x). \tag{3.5.17}$$

Substitution into Eq. (3.5.16) gives

$$C'D = CDCD'' \tag{3.5.18}$$

and

$$\frac{C'(t)}{C^2} = D''(x) = \alpha, \tag{3.5.19}$$

where α is the separation constant. Integrating these differential equations gives the solutions

$$C(t) = \frac{1}{\alpha_1 - \alpha t}, \tag{3.5.20}$$

$$D(x) = \left(\frac{\alpha}{2}\right)x^2 + \beta_1 x + \beta_2, \tag{3.5.21}$$

where $(\alpha_1, \beta_1, \beta_2)$ are arbitrary integration constants. Therefore, Eq. (3.5.16) has the following special rational solution

$$u(x,t) = \frac{\left(\frac{\alpha}{2}\right)x^2 + \beta_1 x + \beta_2}{\alpha_1 - \alpha t}. \tag{3.5.22}$$

A nonstandard explicit finite-difference model for Eq. (3.5.16) is

$$\frac{u_m^{k+1} - u_m^k}{\Delta t} = u_m^{k+1}\left[\frac{u_{m+1}^k - 2u_m^k + u_{m-1}^k}{(\Delta x)^2}\right],$$ (3.5.23)

where the orders of the discrete derivatives are the same as the derivatives of the partial differential equation. Also, the nonlinear term, uu_{xx}, is modeled nonlocally on the lattice, i.e., one term is at time-step k and the other is at $k+1$.

Since we know a special solution of the original partial differential equation, we must require that any best finite difference scheme also have this as a solution. To see whether this is the case for the model of Eq. (3.5.23), let us calculate a special solution by assuming that u_m^k has the form

$$u_m^k = C^k D_m.$$ (3.5.24)

Substitution of Eq. (3.5.24) into Eq. (3.5.23) gives

$$\frac{(C^{k+1} - C^k)D_m}{\Delta t} = C^{k+1}C^k D_m\left[\frac{D_{m+1} - 2D_m + D_{m-1}}{(\Delta x)^2}\right],$$ (3.5.25)

and

$$\frac{C^{k+1} - C^k}{(\Delta t)C^{k+1}C^k} = \frac{D_{m+1} - 2D_m + D_{m-1}}{(\Delta x)^2} = \alpha,$$ (3.5.26)

where α is the separation constant. The two difference equations

$$C^{k+1} - C^k = \alpha(\Delta t)C^{k+1}C^k,$$ (3.5.27)

$$D_{m+1} - 2D_m + D_{m-1} = \alpha(\Delta x)^2,$$ (3.5.28)

have solutions that can be put in the forms [10]

$$C^k = \frac{1}{\alpha_1 - \alpha t_k},$$ (3.5.29)

$$D_m = \left(\frac{\alpha}{2}\right)x_m^2 + \beta_1 x_m + \beta_2,$$ (3.5.30)

where

$$t_k = (\Delta t)k, \qquad x_m = (\Delta x)m, \tag{3.5.31}$$

and $(\alpha_1, \beta_1, \beta_2)$ are arbitrary constants. Thus, a special solution to the partial difference equations given in Eq. (3.5.23) is

$$u_m^k = \frac{(\frac{\alpha}{2})x_m^2 + \beta_1 x_m + \beta_2}{\alpha_1 - \alpha t_k}. \tag{3.5.32}$$

For these special solutions

$$u(x_m, t_k) = u_m^k. \tag{3.5.33}$$

Thus, we conclude that Eq. (3.5.23) is a best finite-difference scheme for Eq. (3.5.16).

It should be acknowledged, however, that another best finite-difference scheme also exists for Eq. (3.5.16) and is given by the implicit scheme

$$\frac{u_m^{k+1} - u_m^k}{\Delta t} + u_m^k \left[\frac{u_{m+1}^{k+1} - 2u_m^{k+1} + u_{m-1}^{k+1}}{(\Delta x)^2} \right] = 0. \tag{3.5.34}$$

By the method of separation-of-variables, it is easily determined that this partial difference equation has the expression of Eq. (3.5.32) as a solution.

At this stage of the discrete modeling process, there is no reason to choose one form of best finite-difference scheme over the other. This result illustrates the (sometimes) ambiguities that arise even in the application of the nonstandard modeling rules. Additional information on the properties of the solutions to the partial differential equation is needed to make a (possible) unique selection.

Throughout the remainder of this book, we will use the five nonstandard modeling rules and possible modifications of them (when needed) to construct discrete models for a variety of special classes of differential equations.

References

1. R. E. Mickens, *Numerical Methods for Partial Differential Equations* 5, 313–325 (1989). Exact solutions to a finite-difference model of a nonlinear reaction-advection equation: Implications for numerical analysis.

2. R. E. Mickens and A. Smith, *Journal of the Franklin Institute* **327**, 143–149 (1990). Finite-difference models of ordinary differential equations: Influence of denominator functions.

3. J. Marsden and A. Weinsten, *Calculus I* (Springer-Verlag, New York, 2nd edition, 1985), section 1.3.

4. R. B. Potts, *American Mathematical Monthly* **89**, 402–407 (1982). Differential and difference equations.

5. V. V. Nemytski and V. V. Stepanov, *Qualitative Theory of Differential Equations* (Princeton University Press; Princeton, NJ; 1969).

6. J. A. Murdock, *Perturbations: Theory and Methods* (Wiley-Interscience, New York, 1991), Appendix D.

7. R. E. Mickens, Difference equation models of differential equations having zero local truncation errors, in *Differential Equations*, I. W. Knowles and R. T. Lewis, editors (North-Holland, Amsterdam, 1984), pp. 445–449.

8. E. Zauderer, *Partial Differential Equations of Applied Mathematics* (Wiley-Interscience, New York, 1983).

9. F. B. Hilderbrand, *Finite-Difference Equations and Simulations* (Prentice-Hall; Englewood Cliffs, NJ; 1968).

10. R. E. Mickens, *Difference Equations: Theory and Applications* (Van Nostrand Reinhold, New York, 1990), section 4.3.

11. R. E. Mickens, *Mathematics and Computer Modelling* **11**, 528–530 (1988). Different equation models of differential equations.

12. A. R. Mitchell and J. C. Bruch, *Numerical Methods for Partial Differential Equations* **1**, 13–23 (1985). A numerical study of chaos in a reaction-diffusion equation.

13. J. Murray, *Lectures on Nonlinear Differential Equation Models in Biology* (Clarendon Press, Oxford, 1977).

14. R. D. Richtmyer and K. W. Morton, *Difference Methods for Initial-Value Problems* (Wiley-Interscience, New York, 2nd edition, 1967).

15. A. R. Mitchell and D. F. Griffiths, *Finite Difference Methods in Partial Differential Equations* (Wiley, New York, 1980).

16. M. K. Jain, *Numerical Solution of Differential Equations* (Halsted Press, New York, 1984).

17. J. J. Stokes, *Nonlinear Vibrations* (Interscience, New York, 1950).

18. R. E. Mickens, O. Oyedeji and C. R. McIntyre, *Journal of Sound and Vibration* **130**, 509–512 (1989). A difference-equation model of the Duffing equation.

19. R. B. Potts, *Journal of the Australian Mathematics Society (Series B)* **23**, 349–356 (1982). Best difference equation approximation to Duffing's equation.

Chapter 4
FIRST ORDER ODE'S

4.1 Introduction

This chapter presents a technique for constructing finite-difference models of a single scalar differential equation. The work to be discussed is based on the investigations of Mickens and Smith [1] and Mickens [2]. The equation to be investigated is the autonomous first-order differential equation

$$\frac{dy}{dt} = f(y). \tag{4.1.1}$$

Our analysis will be done under the assumption that

$$f(y) = 0, \tag{4.1.2}$$

has only simple zeros. Our goal is to construct discrete models of Eq. (4.1.1) that do not exhibit elementary numerical instabilities, i.e., solutions to the finite-difference equation that do not correspond to any of the solutions to the differential equation. For Eq. (4.1.1) numerical instabilities will occur whenever the linear stability properties of any of the fixed-points for the difference scheme differs from that of the differential equation [3, 4, 5].

The main purpose of this chapter is to prove, for Eq. (4.1.1), that it is possible to construct finite-difference schemes that have the correct linear stability properties for all finite step-sizes [1, 2]. The proof involves the explicit construction of such schemes.

This chapter extends the analysis given in the latter half of Section 2.7. In summary, we found there that numerical instabilities occur by several mechanisms: (a) For a central difference scheme, the numerical instabilities are a consequence of the order of the difference scheme being higher than the order of the differential equation. (b) For forward Euler schemes, the numerical instabilities arise when

94

the step-size is larger than some fixed, finite value, $h > h^* > 0$. (c) The implicit backward Euler scheme exhibits super-stability, i.e., its numerical instabilities occur above some threshold value of the step-size, say $h > h_0$, such that all the fixed-points of the difference scheme become stable. (d) The use of higher order (in h) schemes, such as Runge-Kutta methods, gives rise to numerical instabilities because of the appearance of spurious real fixed-points for $h > h_1$.

4.2 A New Finite-Difference Scheme

Denote the fixed-points of Eq. (4.1.1) by

$$\{\bar{y}^{(i)}; i = 1, 2, \ldots, I\}, \tag{4.2.1}$$

where I may be unbounded. The fixed-points are the solutions to the equation

$$f(\bar{y}) = 0. \tag{4.2.2}$$

Define R_i to be

$$R_i \equiv \frac{df[\bar{y}^{(i)}]}{dy}, \tag{4.2.3}$$

and R^* as

$$R^* \equiv \text{Max}\{|R_i|; i = 1, 2, \ldots, I\}. \tag{4.2.4}$$

Linear stability analysis applied to the i-th fixed-point gives the following results [6]:

(i) If $R_i > 0$, the fixed-point $y(t) = \bar{y}^{(i)}$ is linearly unstable.

(ii) If $R_i < 0$, the fixed-point $y(t) = \bar{y}^{(i)}$ is linearly stable.

Now consider the following finite-difference scheme for Eq. (4.1.1)

$$\frac{y_{k+1} - y_k}{\left[\frac{\phi(hR^*)}{R^*}\right]} = f(y_k), \tag{4.2.5}$$

where $\phi(z)$ has the two properties

$$\phi(z) = z + O(z^2), \qquad z \to 0, \tag{4.2.6a}$$

$$0 < \phi(z) < 1, \qquad z > 0. \tag{4.2.6b}$$

Theorem. *The finite-difference scheme of Eq. (4.2.5) has fixed-points with exactly the same linear stability properties as the differential equation*

$$\frac{dy}{dt} = f(y), \tag{4.2.7}$$

for all $h > 0$.

Proof. Represent a perturbation about the i-th fixed-point by

$$y_k = \bar{y}^{(i)} + \epsilon_k. \tag{4.2.8}$$

The linear stability analysis equation for ϵ_k is

$$\frac{\epsilon_{k+1} - \epsilon_k}{\left[\frac{\phi(hR^*)}{R^*}\right]} = R_i \epsilon_k, \tag{4.2.9}$$

or

$$\epsilon_{k+1} = \left[1 + \left(\frac{R_i}{R^*}\right)\phi(hR^*)\right]\epsilon_k, \tag{4.2.10}$$

which has the solution

$$\epsilon_k = \epsilon_0 \left[1 + \left(\frac{R_i}{R^*}\right)\phi(hR^*)\right]^k. \tag{4.2.11}$$

For $R_i > 0$, the fixed-point of the differential equation is linearly unstable. Thus, it follows that

$$1 + \left(\frac{R_i}{R^*}\right)\phi(hR^*) > 0, \qquad h > 0. \tag{4.2.12}$$

Therefore, $y_k = \bar{y}^{(i)}$ is also linearly unstable for $h > 0$.

For $R_i < 0$, the fixed-point of the differential equation is linearly stable. In this instance

$$0 < 1 - \left(\frac{|R_i|}{R^*}\right)\phi(hR) < 1, \qquad h > 0, \tag{4.2.13}$$

a result that follows directly from Eqs. (4.2.4) and (4.2.6). Therefore, $y_k = \bar{y}^{(i)}$ is linearly stable for $h > 0$.

This theorem shows that it is possible to construct discrete models for a single scalar, autonomous ordinary differential equation such that no elementary numerical instabilities occur in their solutions. This result is related to the fact that most elementary numerical instabilities arise from a given fixed-point having the opposite linear stability properties in the difference scheme and the differential equation. The above construction demonstrates that to achieve the correct linear stability behavior, a generalized definition of the discrete derivative must be used [1]. None of the standard finite-difference modeling procedures have this property, namely, the correct linear stability behavior for all step-sizes.

The above finite-difference scheme uses the following *denominator function* for the discrete first-derivative

$$D(h, R^*) \equiv \frac{\phi(hR^*)}{R^*}, \tag{4.2.14}$$

where ϕ and R^* are given by Eqs. (4.2.4) and (4.2.6). This form replaces the simple "h" function found in the standard finite-difference schemes, i.e.,

$$\frac{dy}{dt} \to \frac{y_{k+1} - y_k}{h}, \qquad \text{(standard schemes)}. \tag{4.2.15}$$

Note that in the limits ($h \to 0$, $k \to \infty$, $hk = t =$ fixed), the generalized discrete derivative reduces to the first derivative, i.e.,

$$\lim_{\substack{h \to 0 \\ k \to \infty \\ hk = t = \text{fixed}}} \frac{y_{k+1} - y_k}{\left[\frac{\phi(hR^*)}{R^*}\right]} = \frac{dy}{dt}. \tag{4.2.16}$$

However, for fixed $h > 0$ and a given value of k, the generalized discrete derivative may have a numerical magnitude that differs greatly from the standard discrete derivatives such as those given by the central difference, forward Euler and backward

Euler representations. The denominator function $D(h, R^*)$ can be considered a renormalization of the step-size h to a new value h', i.e.,

$$h \rightarrow h' = D(h, R^*). \qquad (4.2.17)$$

This concept of renormalized variables and constants occurs frequently in various areas of the sciences [6].

It should he noted that, except for the requirements given in Eqs. (4.2.6), the function $\phi(z)$ can be arbitrary. However, a particularly useful and simple functional form for $\phi(z)$ is the following expression which occurs in many exact finite-difference schemes

$$\phi(z) = 1 - \epsilon^{-z}. \qquad (4.2.18)$$

Finally, it should be observed that the difference scheme of Eq. (4.2.5) resolves the problem of "super-stability" that occurs in the use of the backward Euler scheme [3, 4, 7, 8, 9]; see Section 2.7. The finite-difference scheme, for this case, is

$$\frac{y_{k+1} - y_k}{\left[\frac{\phi(hR^*)}{R^*}\right]} = f(y_{k+1}). \qquad (4.2.19)$$

For a perturbation about the i-th fixed-point

$$y_k = \bar{y}^{(i)} + \epsilon_k, \qquad (4.2.20)$$

the equation for ϵ_k is

$$\frac{\epsilon_{k+1} - \epsilon_k}{\left[\frac{\phi(hR^*)}{R^*}\right]} = R_i \epsilon_{k+1}, \qquad (4.2.21)$$

which has the solution

$$\epsilon_k = \epsilon_0 \left[\frac{1}{1 - \left(\frac{R_i}{R^*}\right)\phi(hR^*)}\right]^k. \qquad (4.2.22)$$

If $R_i < 0$, the fixed-point for the differential equation is linearly stable. Now, for this case

$$1 - \left(\frac{R_i}{R^*}\right)\phi(hR^*) = 1 + \left(\frac{|R_i|}{R^*}\right)\phi(hR^*) > 0. \qquad (4.2.23)$$

This implies that $y_k = \bar{y}^{(i)}$ is also linearly stable for $h > 0$. Likewise, for $R_i > 0$, the fixed-point for the differential equation is linearly unstable. Since

$$0 < \frac{R_i}{R^*} < 1; \qquad 0 < \phi < 1, \ 0 < h < \infty; \tag{4.2.24}$$

it follows that

$$0 < 1 - \left(\frac{R_i}{R^*}\right)\phi(hR^*) < 1, \tag{4.2.25}$$

and it can be concluded that $y_k = \bar{y}^{(i)}$ is linearly unstable. Thus, the finite-difference scheme of Eq. (4.2.19) has fixed-points with exactly the same linear stability properties as the scalar differential equation. This result holds for all step-sizes, $h > 0$. Consequently, "super-stability" will not occur in the discrete model of Eq. (4.2.19).

4.3 Examples

We illustrate the power of the new finite-difference scheme for scalar first-order ordinary differential equations by applying it to three equations: the decay equation, the Logistic differential equation and an equation having three fixed-points.

4.3.1 Decay Equation

The decay differential equation is

$$\frac{dy}{dt} = -\lambda y, \tag{4.3.1}$$

where λ is a positive constant. The function $f(y)$ is

$$f(y) = -\lambda y. \tag{4.3.2}$$

There is a single globally stable fixed-point at $\bar{y}^{(i)} = 0$. In addition,

$$R_1 = -\lambda, \qquad R^* = \lambda. \tag{4.3.3}$$

Now, select for $\phi(z)$, the expression

$$\phi(z) = 1 - e^{-z}. \tag{4.3.4}$$

Substitution of these results into Eq. (4.2.5) gives

$$\frac{y_{k+1} - y_k}{\left(\frac{1-e^{-\lambda h}}{\lambda}\right)} = -\lambda y_k, \qquad (4.3.5)$$

which is the exact finite-difference scheme for the decay equation; see Eq. (3.3.9).

Thus, in this case, the new finite-difference scheme gives the exact discrete model.

4.3.2 Logistic Equation

For the Logistic differential equation

$$\frac{dy}{dt} = y(1-y), \qquad (4.3.6)$$

we have

$$f(y) = y(1-y), \qquad (4.3.7)$$

two fixed-points at

$$\bar{y}^{(1)} = 0, \qquad \bar{y}^{(2)} = 1, \qquad (4.3.8)$$

and

$$R_1 = 1, \qquad R_2 = -1, \qquad R^* = 1. \qquad (4.3.9)$$

The substitution of Eqs. (4.3.7), (4.3.9) and (4.3.4) into Eq. (4.2.5) gives

$$\frac{y_{k+1} - y_k}{1 - e^{-h}} = -y_k(1 - y_k). \qquad (4.3.10)$$

Figure 4.3.1 gives the numerical solution of the Logistic differential equation using Eq. (4.3.10). The initial condition is $y(0) = y_0 = 0.5$ and the step-sizes used in the computation are $h = (0.01, 1.5, 2.5)$. Note that for all three step-sizes, the numerical solution has the same qualitative behavior as the exact solution, i.e., a monotonic increase to the value one.

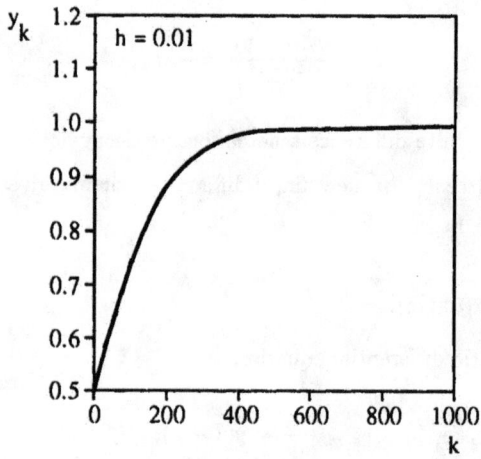

(a)

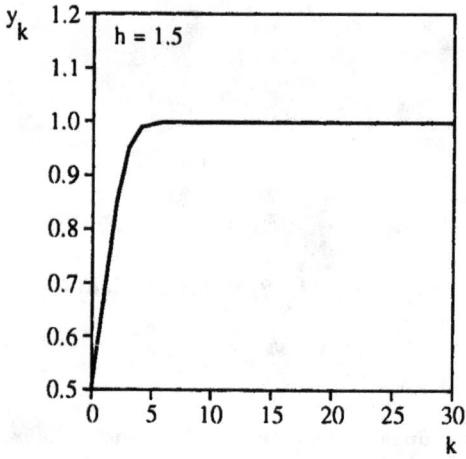

(b)

Figure 4.3.1. Plots of Eq. (4.3.10). For each graph
$y_0 = 0.5$, (a) $h = 0.01$, (b) $h = 1.5$.

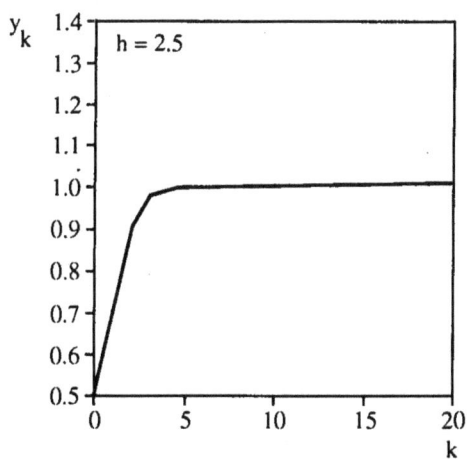

Figure 4.3.1. Plots of Eq. (4.3.10). For each graph
$y_0 = 0.5$, (c) h = 2.5.

4.3.3 ODE with Three Fixed-Points

The simplest ordinary differential equation with three fixed-points is

$$\frac{dy}{dt} = y(1 - y^2). \tag{4.3.11}$$

For this equation

$$f(y) = y(1 - y^2), \tag{4.3.12}$$

$$\bar{y}^{(1)} = 0, \qquad \bar{y}^{(2)} = 1, \qquad \bar{y}^{(3)} = -1, \tag{4.3.13}$$

$$R_1 = 1, \qquad R_2 = R_3 = -2, \qquad R^* = 2. \tag{4.3.14}$$

Using $\phi(z)$ from Eq. (4.3.4), we obtain, on substitution of these results into Eq. (4.2.5), the following discrete model for Eq. (4.3.11)

$$\frac{y_{k+1} - y_k}{\left(\frac{1 - e^{-2h}}{2}\right)} = y_k(1 - y_k^2). \tag{4.3.15}$$

Figure 4.3.2 gives the general behavior of the solutions for various initial values, $y(0) = y_0$. The $(\pm)$ sign denotes the regions where the derivative has a constant sign; at the fixed-point, the derivative must be zero. For $y_0 > 0$, all solutions approach the stable fixed-point at $\bar{y}^{(2)} = 1$. Likewise, for $y_0 < 0$, all solutions approach the other stable fixed-point at $\bar{y}^{(3)} = -1$.

Figure 4.3.3 presents numerical solutions obtained from Eq. (4.3.15). Each graph starts with the initial condition $y(0) = y_0 = 0.5$. The four step-sizes used are $h = (0.01, 0.75, 1.5, 2.5)$. Observe that for all four step-sizes, the numerical functions have exactly the same qualitative behavior as the corresponding solution of the differential equation, i.e., a monotonic increase from $y_0 = 0.5$ to $y_\infty = \bar{y}^{(2)} = 1$.

For purposes of comparison, we consider the standard forward Euler scheme for Eq. (4.3.11); it is

$$\frac{y_{k+1} - y_k}{h} = y_k(1 - y_k^2) \tag{4.3.16}$$

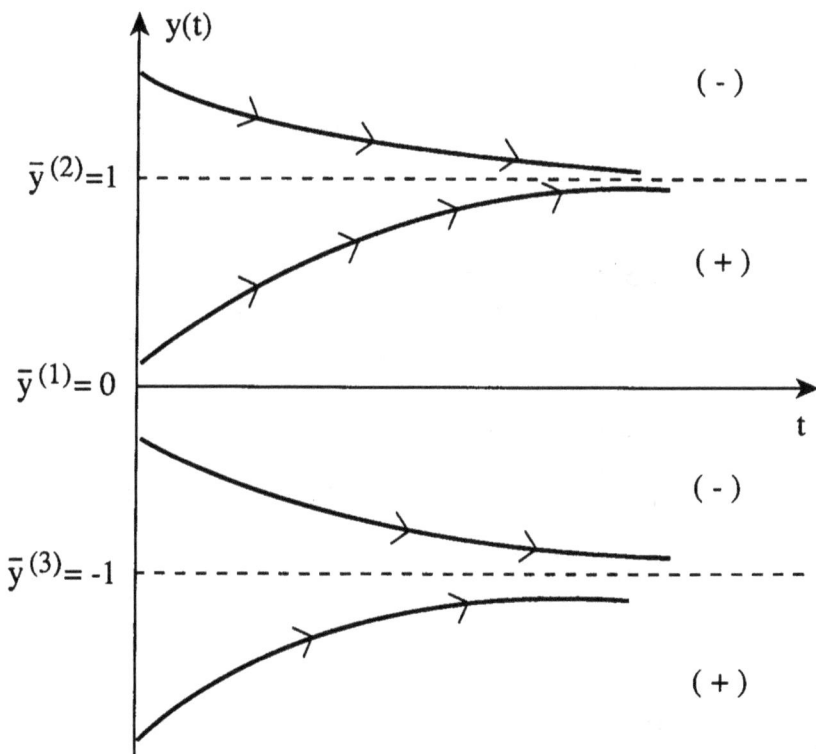

Figure 4.3.2. General solution behavior for Eq. (4.3.11).

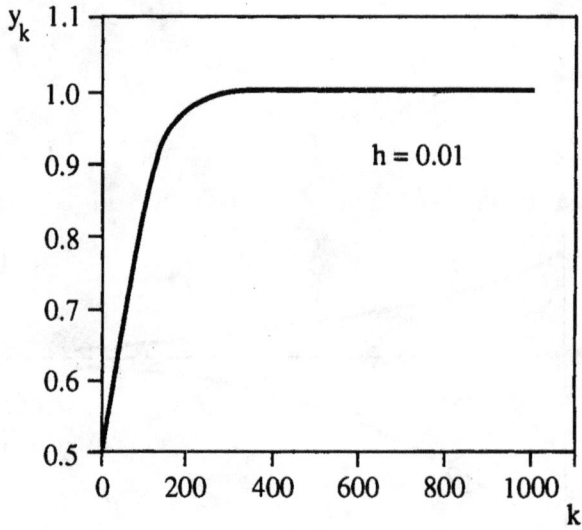

(a)

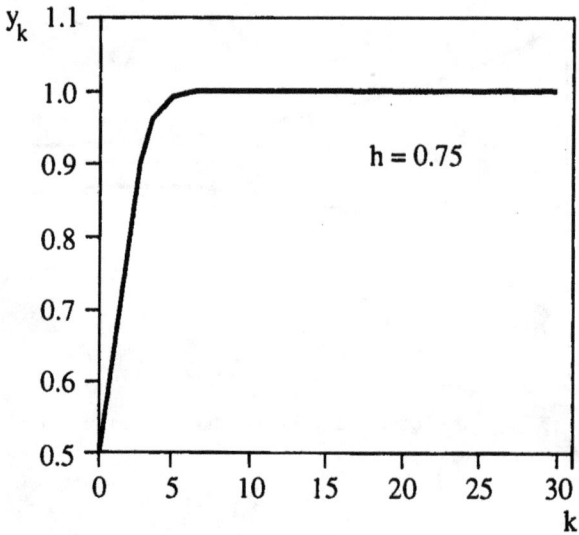

(b)

Figure 4.3.3. Plots of Eq. (4.3.15). For each graph
$y_0 = 0.5$, (a) h = 0.01, (b) h = 0.75.

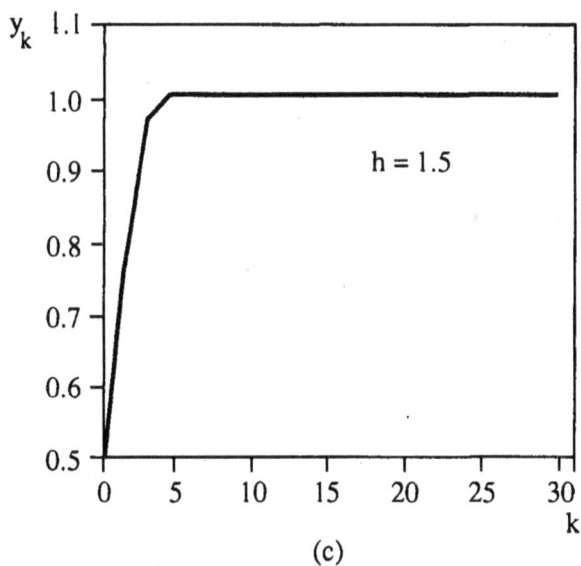

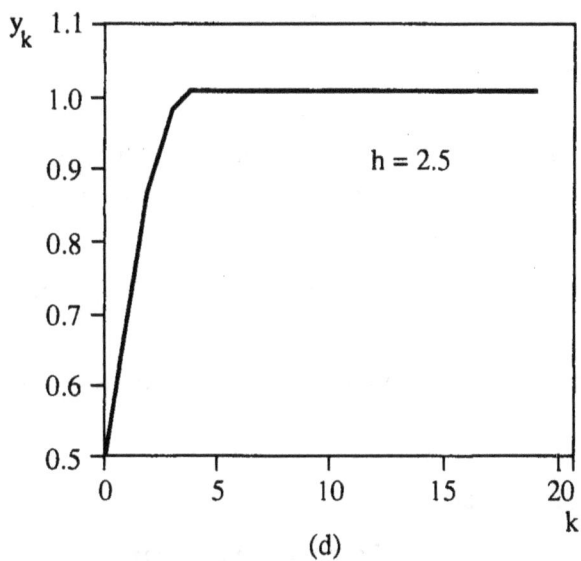

Figure 4.3.3. Plots of Eq. (4.3.15). For each graph
$y_0 = 0.5$, (c) $h = 1.5$, (d) $h = 2.5$.

Perturbations about the three fixed-points

$$y_k = \begin{cases} 1 + \eta_k \\ 0 + \epsilon_k \\ -1 + \eta_k, \end{cases} \qquad (4.3.17)$$

give the following linear stability equations

$$\epsilon_{k+1} = (1 + h)\epsilon_k, \qquad (4.3.18)$$

$$\eta_{k+1} = (1 - 2h)\eta_k. \qquad (4.3.19)$$

From Eq. (4.3.18), it follows that $\bar{y}^{(1)} = 0$ is linearly unstable for all $h > 0$. However, the fixed-points at $\bar{y}^{(2)} = 1$, $\bar{y}^{(3)} = -1$, have the following linear stability properties:

(i) For $0 < h < 0.5$, both fixed-points are linearly stable.

(ii) For $0.5 < h < 1$, both fixed-points are linearly stable; however, the perturbations decrease to zero with an oscillating amplitude.

(iii) For $h > 1$, the two-fixed-points are linearly unstable.

The results given in Figure 4.3.4 are numerical solutions obtained from the forward Euler scheme of Eq. (4.3.16). For each, the initial condition is $y(0) = y_0 = 0.5$ and the respective step-sizes are $h = (0.01, 0.75, 1.5, 2.0)$. Note that the graphs are fully consistent with the above linear stability analysis.

The results of this section can be summarized in the statement that the new finite-difference scheme of Eq. (4.2.5) provides superior discrete models of the three differential equations studied as compared to the use of the standard forward Euler scheme.

4.4 Nonstandard Schemes

Chapter 3 provided a set of nonstandard modeling rules. We now apply them to two of the differential equations examined in the last section. The new modeling rule, to be added to the results of Section 4.2, is the requirement that nonlinear terms be modeled nonlocally on the computational grid.

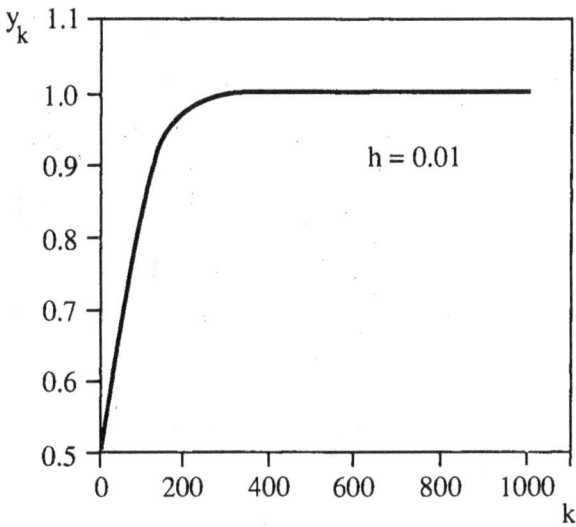

(a)

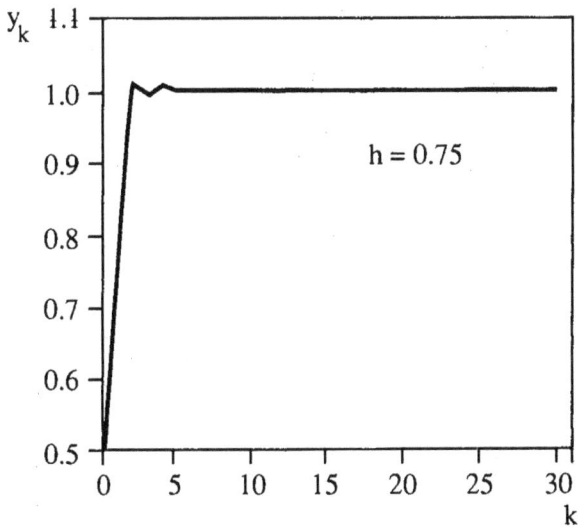

(b)

Figure 4.3.4. Plots of Eq. (4.3.16). For each graph
$y_0 = 0.5$, (a) $h = 0.01$, (b) $h = 0.75$.

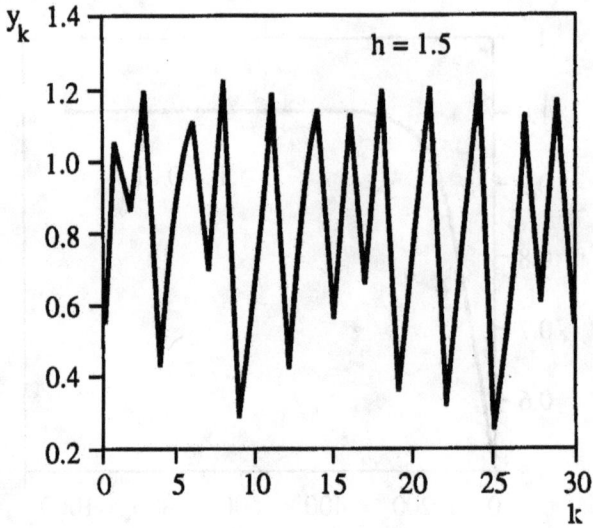

(c)

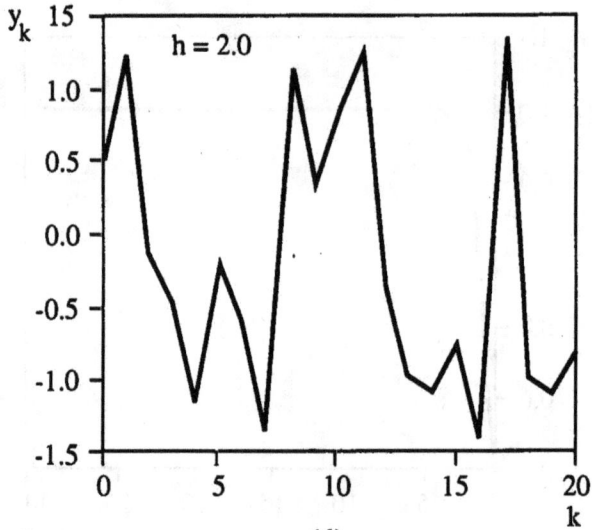

(d)

Figure 4.3.4. Plots of Eq. (4.3.16). For each graph
$y_0 = 0.5$, (c) $h = 1.5$, (d) $h = 2.0$.

4.4.1 Logistic Equation

The discrete scheme for the Logistic differential equation, with a nonlocal nonlinear term, is

$$\frac{y_{k+1} - y_k}{1 - e^{-h}} = y_k(1 - y_{k+1}). \tag{4.4.1}$$

This difference equation can be solved exactly by use of the transformation [10]

$$y_k = \frac{1}{w_k}. \tag{4.4.2}$$

This gives

$$w_{k+1} - \left(\frac{1}{2 - e^{-h}}\right) w_k = \frac{1 - e^{-h}}{2 - e^{-h}}, \tag{4.4.3}$$

whose exact solution is [10]

$$w_k = 1 + A(2 - e^{-h})^{-k}, \tag{4.4.4}$$

where A is an arbitrary constant. Imposing the initial condition, $y(0) = y_0$, we obtain

$$y_k = \frac{y_0}{y_0 + (1 - y_0)(2 - e^{-h})^{-k}}. \tag{4.4.5}$$

Note that

$$1 < 2 - e^{-h} < 2, \qquad h > 0, \tag{4.4.6}$$

consequently,

$$g_k = (2 - e^{-h})^{-k}, \tag{4.4.7}$$

is an exponentially decreasing function of k. Examination of Eq. (4.4.5) shows that all the solutions of Eq. (4.4.1) have the same qualitative properties as the solutions to the Logistic differential equation for all step-sizes, $h > 0$.

4.4.2 ODE with Three Fixed-Points

A discrete model for Eq. (4.3.11), with a nonlocal nonlinear term, is

$$\frac{y_{k+1} - y_k}{\left(\frac{1 - e^{-2h}}{2}\right)} = y_k(1 - y_{k+1}y_k). \tag{4.4.8}$$

This expression is linear in y_{k+1}; solving for it gives

$$y_{k+1} = \frac{(1 + \bar{\phi})y_k}{1 + \bar{\phi}y_k^2},$$ (4.4.9a)

where

$$\bar{\phi} = \frac{1 - \epsilon^{-2h}}{2}.$$ (4.4.9b)

Numerical solutions of Eq. (4.4.8) or (4.4.9) are plotted in Figure 4.4.1 for $y_0 = 0.5$ and the step-sizes $h = (0.01, 0.75, 1.5, 2.5)$. Observe that for all the selected step-sizes, the numerical solutions increase monotonically toward the limiting value of $y_\infty = \bar{y}^{(2)} = 1$. This is exactly the same qualitative behavior as the corresponding solution to the differential equation.

For purposes of comparison, it is of interest to also examine the numerical solutions of the discrete model

$$\frac{y_{k+1} - y_k}{h} = y_k(1 - y_{k+1}y_k).$$ (4.4.10)

This model is constructed by using a standard forward Euler scheme for the first-derivative and a nonlocal representation for the nonlinear term. Solving for y_k gives

$$y_{k+1} = \frac{(1 + h)y_k}{1 + hy_k^2}.$$ (4.4.11)

Figure 4.4.2 presents numerical solutions to the finite-difference scheme of Eq. (4.4.10) or (4.4.11). The initial condition and step-size values are the same as in Figure 4.4.1. The obtained results can be explained by means of a linear stability analysis.

Perturbations about the three fixed-points of Eq. (4.4.10) give the following linear stability equations

$$\epsilon_{k+1} = (1 + h)\epsilon_k,$$ (4.4.12)

$$\eta_{k+1} = \left(\frac{1 - h}{1 + h}\right)\eta_k.$$ (4.4.13)

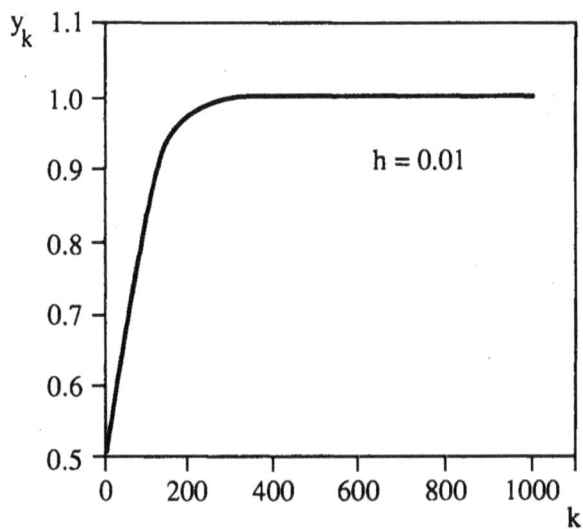

(a)

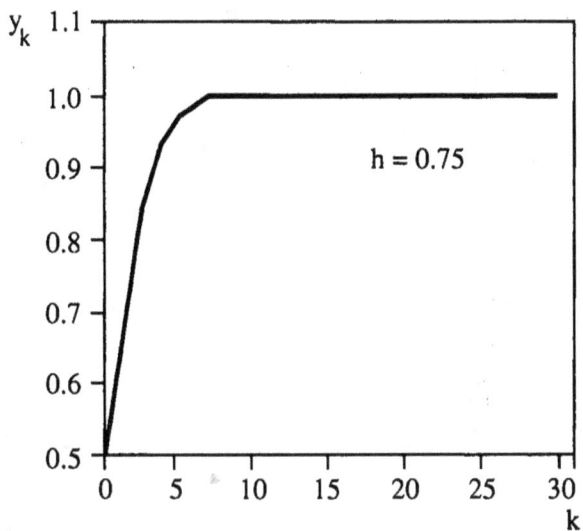

(b)

Figure 4.4.1. Plots of Eq. (4.4.8). For each graph
$y_0 = 0.5$, (a) $h = 0.01$, (b) $h = 0.75$.

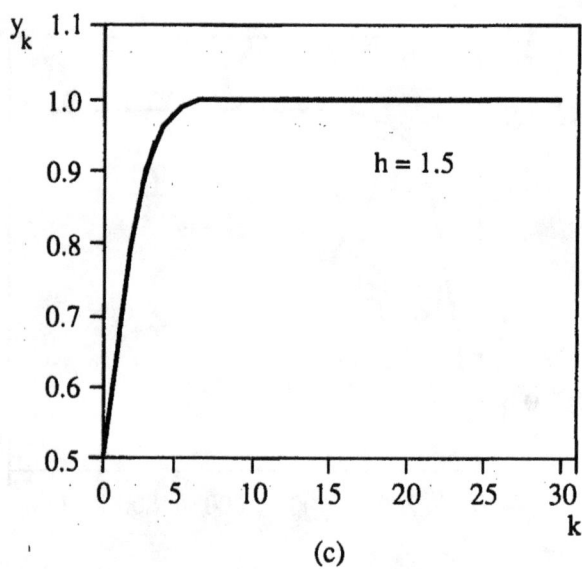

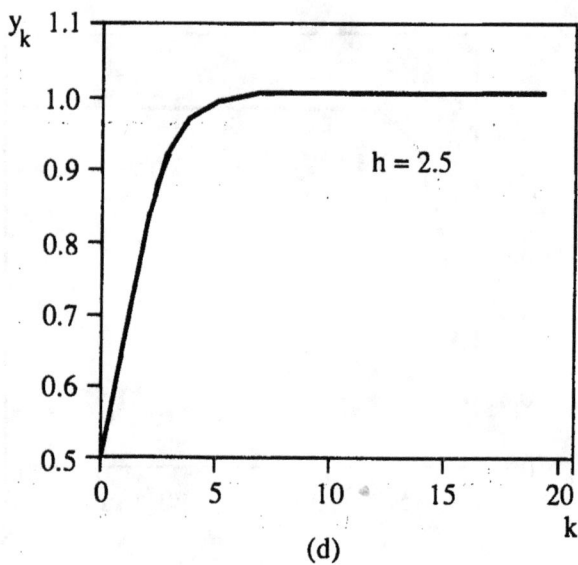

Figure 4.4.1. Plots of Eq. (4.4.8). For each graph
$y_0 = 0.5$, (c) $h = 1.5$, (d) $h = 2.5$.

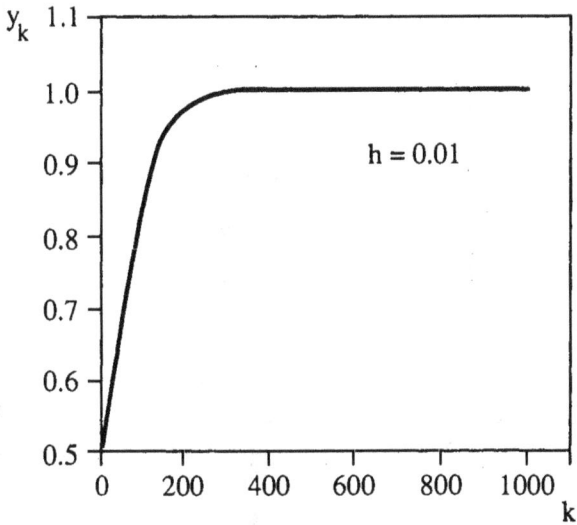

(a)

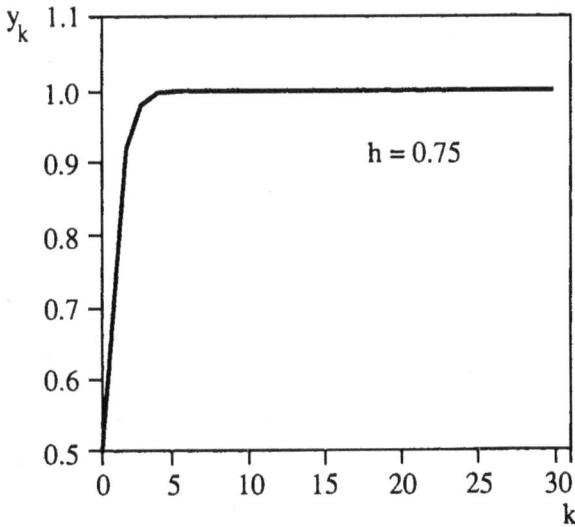

(b)

Figure 4.4.2. Plots of Eq. (4.4.10). For each graph
$y_0 = 0.5$, (a) $h = 0.01$, (b) $h = 0.75$.

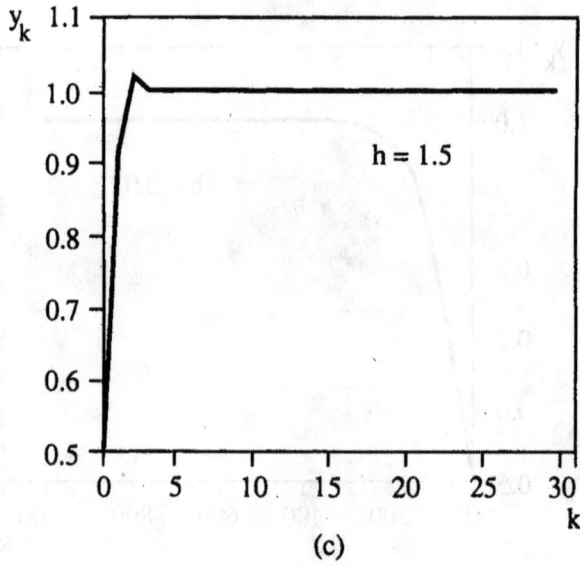

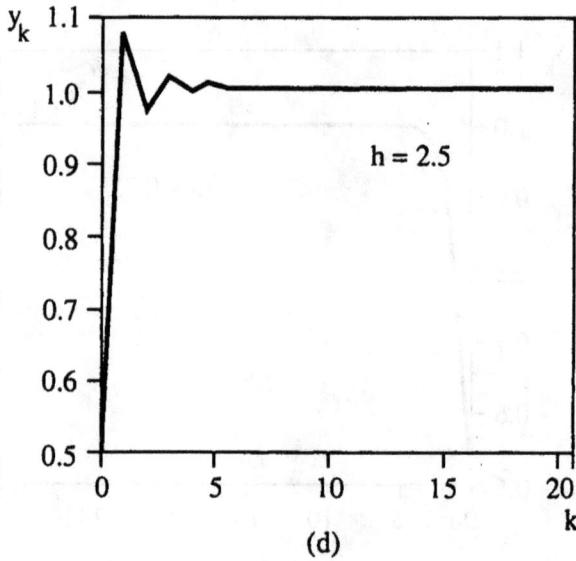

Figure 4.4.2. Plots of Eq. (4.4.10). For each graph
$y_0 = 0.5$, (c) $h = 1.5$, (d) $h = 2.5$.

(See Eq. (4.3.17).) From Eq. (4.4.12), it can be concluded that the fixed-point at $y_k = \bar{y}^{(1)} = 0$ is unstable for all step-sizes, $h > 0$. However, the two other fixed-points, at $\bar{y}^{(2)} = 1$ and $\bar{y}^{(3)} = -1$, have the following properties:

(i) For $0 < h < 1$, the fixed-points are linearly stable.

(ii) For $h > 1$, the fixed-points are linearly stable; but, the perturbations decrease with an oscillating amplitude.

This is just the behavior seen in the various graphs of Figure 4.4.2.

4.5 Discussion

The calculations presented in this chapter show, for a scalar ordinary differential equation

$$\frac{dy}{dt} = f(y), \tag{4.5.1}$$

that the use of a renormalized denominator function

$$h \to \frac{1 - e^{-R^* h}}{R^*}, \tag{4.5.2}$$

leads to discrete models for which the fixed-points have the correct linear stability properties for all step-sizes, $h > 0$. This result is obtained whether or not a local or a nonlocal representation is used for the function $f(y)$. The procedure given for these constructions is the simplest possible for the differential equations investigated. However, more complicated discrete models exist. For example, consider the differential equation with three fixed-points

$$\frac{dy}{dt} = y(1 + y)(1 - y). \tag{4.5.3}$$

A finite-difference scheme that incorporates the maximum symmetry in the nonlocal modeling of the nonlinear term is

$$\left[\frac{\Delta y_k}{\phi} - y_{k+1}(1 + y_k)(1 - y_k)\right] + \left[\frac{\Delta y_k}{\phi} - y_k(1 + y_{k+1})(1 - y_k)\right]$$
$$+ \left[\frac{\Delta y_k}{\phi} - y_k(1 + y_k)(1 - y_{k+1})\right]$$
$$- \left[\frac{\Delta y_k}{\phi} - y_k(1 + y_k)(1 - y_k)\right] = 0, \tag{4.5.4}$$

where

$$\Delta y_k = y_{k+1} - y_k, \qquad \phi = \frac{1 - e^{-2h}}{2}. \tag{4.5.5}$$

(Such a form has been investigated by Price et al. [11] for an ordinary differential equation similar in form to Eq. (4.5.3). However, they consider the case where $\phi = h$.) This equation can be solved for y_{k+1} to give

$$y_{k+1} = \frac{\left[(5 - e^{-2h}) + (1 - e^{-2h})y_k^2\right] y_k}{(3 + e^{-2h}) + 3(1 - e^{-2h})y_k^2}. \tag{4.5.6}$$

A geometrical analysis [10] of Eq. (4.5.6) shows that if $y_0 > 0$, then y_k converges monotonically to the fixed-point at $\bar{y}^{(2)} = +1$. Similarly, if $y_0 < 0$, then y_k converges monotonically to the fixed-point at $\bar{y}^{(3)} = -1$. This result holds true for all $h > 0$ and corresponds exactly to the qualitative behavior of the various solutions to the differential equation. See Figures 4.5.1 and 4.5.2.

Finally, it should be emphasized that these calculations indicate that the use of a renormalized denominator function has a more important effect on the solution behavior of a discrete model than does the use of a nonlocal representation for the nonlinear term. Of course, putting both in the same discrete model produces better results.

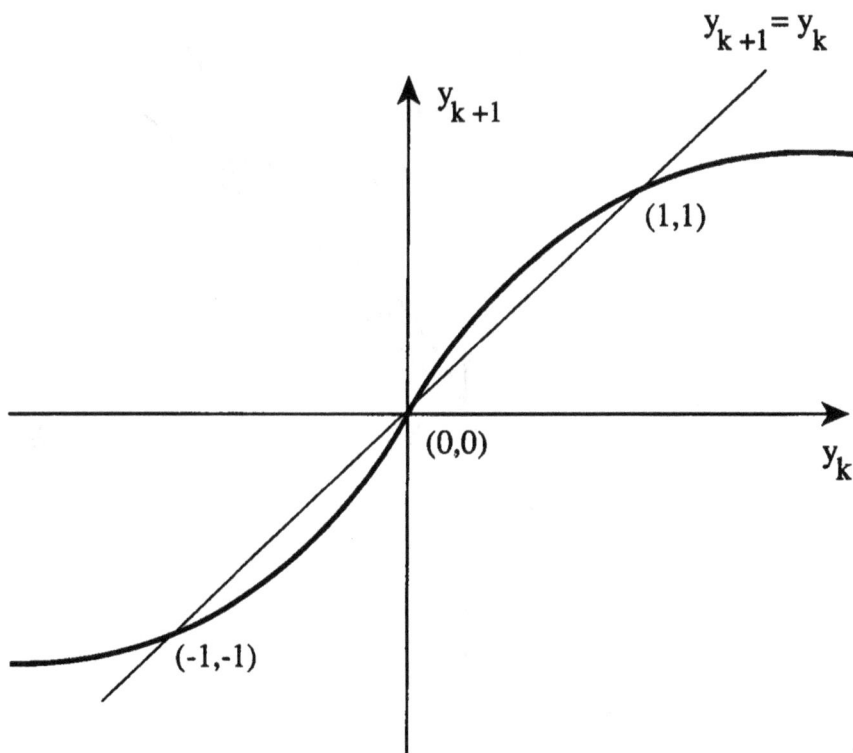

Figure 4.5.1. Plot of Eq. (4.5.6). The fixed-points
are located at (-1,-1), (0,0) and (1,1).

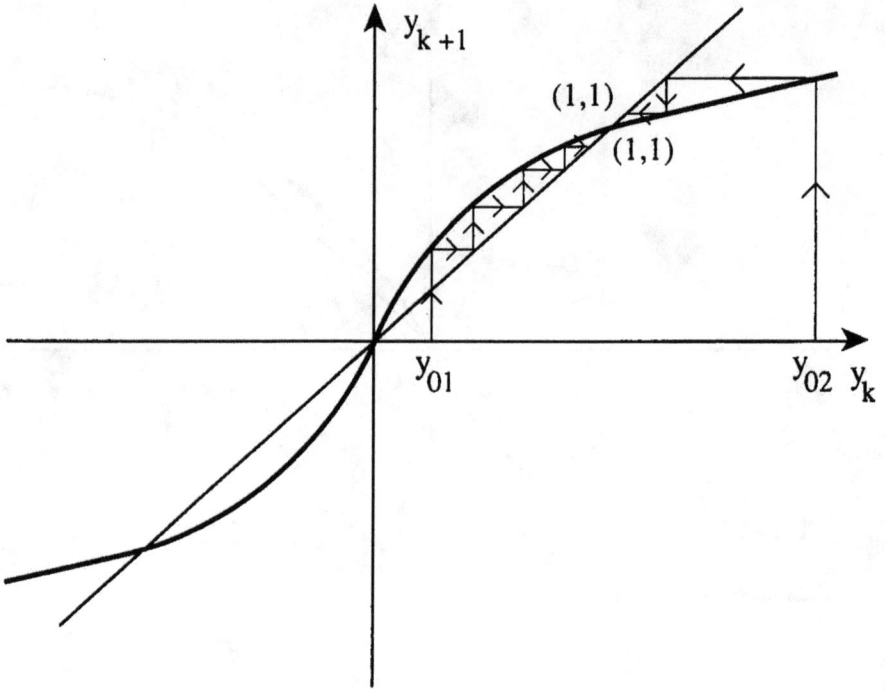

Figure 4.5.2. Typical trajectories for Eq. (4.5.6) with
$y_0 > 0$: $0 < y_{01} < 1$ and $1 < y_{02}$.

References

1. R. E. Mickens and A. Smith, *Journal of the Franklin Institute* **327**, 143–149 (1990). Finite-difference models of ordinary differential equations: Influence of denominator functions.

2. R. E. Mickens, *Dynamic Systems and Applications* **1**, 329–340 (1992). Finite-difference schemes having the correct linear stability properties for all finite step-sizes II.

3. R. M. Corless, C. Essex and M. A. H. Nerenberg, *Physics Letters* **A157**, 27–36 (1991). Numerical methods can suppress chaos.

4. A. Iserles, A. T. Peplow and A. M. Stuart, *SIAM Journal of Numerical Analysis* **28**, 1723–1751 (1991). A unified approach to spurious solutions introduced by time discretization.

5. R. E. Mickens, Runge-Kutta schemes and numerical instabilities: The Logistic equation, in *Differential Equations and Mathematical Physics*, I. Knowles and Y. Saito, editors (Springer-Verlag, Berlin, 1987), pp. 337–341.

6. E. R. Caianiello, *Combinatorics and Renormalization in Quantum Field Theory* (W. A. Benjamin; Reading, MA; 1973).

7. G. Dahlquist, L. Edsberg, G. Skollermo and G. Soderlind, Are the numerical methods and software satisfactory for chemical kinetics?, in *Numerical Integration of Differential Equations and Large Linear Systems*, J. Hinze, editor (Springer-Verlag, Berlin, 1982), pp. 149–164.

8. L. Dieci and D. Estep, Georgia Institute of Technology, Tech. Rep. Math. 050290-039 (1990). Some stability aspects of schemes for the adaptive integration of stiff initial value problems.

9. E. N. Lorenz, *Physica* **D35**, 299–317 (1989). Computational chaos — A prelude to computational instability.

10. R. E. Mickens, *Difference Equations: Theory and Applications* (Van Nostrand Reinhold, New York, 2nd edition, 1990).

11. W. G. Price, Y. Wang and E. H. Twizell, *Numerical Methods for Partial Differential Equations* **9**, 213–224 (1993). A second-order, chaos-free, explicit method for the numerical solution of a cubic reaction problem in neurophysiology.

Chapter 5

SECOND-ORDER, NONLINEAR OSCILLATOR EQUATIONS

5.1 Introduction

Within the context of traditional classical mechanics, a one-dimensional conservative oscillator is described by a differential equation having the form [1–4]

$$\frac{d^2y}{dt^2} + f(y) = 0. \tag{5.1.1}$$

This equation has a first-integral

$$\left(\frac{1}{2}\right)\left(\frac{dy}{dt}\right)^2 + V(y) = E = \text{constant}, \tag{5.1.2}$$

where E is the constant total energy and $V(y)$, given by

$$V(y) = \int f(y)dy, \tag{5.1.3}$$

is the potential energy. Examples of several important conservative oscillator equations are the Duffing equation [3–5]

$$\frac{d^2y}{dt^2} + y + \epsilon y^3 = 0, \tag{5.1.4}$$

the modified Duffing equation

$$\frac{d^2y}{dt^2} + y^3 = 0, \tag{5.1.5}$$

and an equation arising in laser physics [6]

$$\frac{d^2y}{dt^2} + y + \frac{\epsilon y}{1 + \lambda y^2} = 0. \tag{5.1.6}$$

In these equations, ϵ and λ are generally positive constants.

A nonlinear oscillator is defined to be conservative if the differential equation is invariant under the time transformation

$$t \to -t. \tag{5.1.7}$$

The oscillator of Eq. (5.1.1) certainly has this form. However, a larger class of nonlinear conservative oscillators can be considered. For example, the equation of motion of a particle on a rotating parabola is [5]

$$\frac{d^2 y}{dt^2} + \left[\frac{1 + \epsilon(\dot{y})^2}{1 + \epsilon y^2}\right] y = 0, \tag{5.1.8}$$

where $\dot{y} = dy/dt$. This latter differential equation is also invariant under the transformation of Eq. (5.1.7). Consequently, we are led to consider a generalized conservative oscillator differential equation that has the structure

$$\frac{d^2 y}{dt^2} + g[y^2, (\dot{y})^2] y = 0. \tag{5.1.9}$$

Another important class of one-dimensional oscillators is those that have limit-cycles. In general, these systems asymptotically go to a well defined periodic state. The particular periodic state that the system finds itself in may depend on the initial conditions, but, the properties of the various periodic states are functions only of the system parameters [2–5]. The prime example of such a system is the van der Pol oscillator [2, 3, 5]

$$\frac{d^2 y}{dt^2} + y = \epsilon(1 - y^2)\frac{dy}{dt}, \qquad \epsilon > 0. \tag{5.1.10}$$

This differential equation has a unique periodic solution that can be reached from any set of initial conditions in the finite $(y, \dot{y})$ phase-plane.

The purpose of this chapter is to investigate the mathematical properties of various discrete models for both conservative and limit-cycle one-dimensional oscillator differential equations. The particular equations studied will be the Duffing and van der Pol differential equations. The techniques used to construct the finite-difference

122

schemes will be based on the nonstandard modeling rules of Chapter 3. The major mathematical procedure that will be applied to obtain the analytical properties of the solutions to the difference equations is a discrete variable perturbation method formulated by Mickens [7, 8].

The books of Greenspan [9, 10] provide a good summary of his work, as well as the research of others, on the general topic of discrete physical models. Other related work on discrefe-time Hamiltonian dynamics appears in references [11–14].

5.2 Mathematical Preliminaries

Consider the following class of nonlinear, second-order difference equations [8, 9]

$$\Gamma y_k = \epsilon f(y_{k+1}, y_k, y_{k-1}), \tag{5.2.1}$$

where ϵ is a parameter satisfying the condition

$$0 < \epsilon \ll 1, \tag{5.2.2}$$

and the operator Γ is defined by the relation

$$\Gamma y_k \equiv \frac{y_{k+1} - 2y_k + y_{k-1}}{4\sin^2\left(\frac{h}{2}\right)} + y_k. \tag{5.2.3}$$

We now construct a multi-discrete-variable procedure [9] to obtain perturbation solutions to Eq. (5.2.1). We begin by introducing two discrete variables k and $s = \epsilon k$ and assume that the solution to Eq. (5.2.1) has the form

$$y_k \equiv y(k, s, \epsilon) = y_0(k, s) + \epsilon y_1(k, s) + O(\epsilon^2), \tag{5.2.4}$$

where y_k is assumed to have at least a first partial derivative with respect to s. On the basis of these assumptions, we have

$$y_{k+1} = y(k+1, s+\epsilon, \epsilon) = y_0(k+1, s+\epsilon) + \epsilon y_1(k+1, s+\epsilon) + O(\epsilon^2), \tag{5.2.5}$$

$$y(k+1, s+\epsilon) = y_0(k+1, s) + \epsilon \frac{\partial y_0(k+1, s)}{\partial s} + O(\epsilon^2), \qquad (5.2.6)$$

$$y_1(k+1, s+\epsilon) = y_1(k+1, s) + O(\epsilon), \qquad (5.2.7)$$

and

$$y_{k+1} = y_0(k+1, s) + \epsilon \left[y_1(k+1, s) + \frac{\partial y_0(k+1, s)}{\partial s} \right] + O(\epsilon^2), \qquad (5.2.8)$$

$$y_{k-1} = y_0(k-1, s) + \epsilon \left[y_1(k-1, s) - \frac{\partial y_0(k-1, s)}{\partial s} \right] + O(\epsilon^2). \qquad (5.2.9)$$

Substituting Eqs. (5.2.4), (5.2.8) and (5.2.9) into Eq. (5.2.1), and setting the coefficients of the ϵ^0 and ϵ^1 terms equal to zero, gives the following determining equations for the unknown functions $y_0(k, s)$ and $y_1(k, s)$:

$$\Gamma y_0(k, s) = 0, \qquad (5.2.10)$$

$$\Gamma y_1(k, s) = \frac{1}{4 \sin^2 \left(\frac{h}{2} \right)} \left[\frac{\partial y_0(k-1, s)}{\partial s} - \frac{\partial y_0(k+1, s)}{\partial s} \right]$$
$$+ f[y_0(k+1, s), y_0(k, s), y_0(k-1, s)]. \qquad (5.2.11)$$

The first equation has the general solution

$$y_0(k, s) = A(s) \cos(hk) + B(s) \sin(hk), \qquad (5.2.12)$$

where, at present, $A(s)$ and $B(s)$ are unknown functions.

If Eq. (5.2.12) is substituted into the right-side of Eq. (5.2.11), then the following result is obtained

$$\Gamma y_1(k, s) = \left[\lambda \frac{dA}{ds} + M_1(A, B, h) \right] \sin(hk) + \left[-\lambda \frac{dB}{ds} + N_1(A, B, h) \right] \cos(hk)$$
$$+ \text{(higher-order harmonies)}, \qquad (5.2.13)$$

where

$$\lambda = \frac{\sin(h)}{2 \sin^2 \left(\frac{h}{2} \right)}, \qquad (5.2.14)$$

and M_1 and N_1 are obtained from the Fourier series expansion of the function

$$f[y_0(k+1,s), y_0(k,s), y_0(k-1,s)] = \sum_{\ell=1}^{\infty}[M_\ell \sin(\ell hk) + N_\ell \cos(\ell hk)]. \quad (5.2.15)$$

The "higher-order harmonics" term in Eq. (5.2.13) is the sum on the right-side of Eq. (5.2.15) for $\ell \geq 2$. If $y_1(k,s)$ is to be bounded, i.e., contain no secular terms, then the coefficients of the $\sin(hk)$ and $\cos(hk)$ terms must be zero [8]. This condition gives equations that can be solved to get the functions $A(s)$ and $B(s)$; they are

$$\lambda \frac{dA}{ds} + M_1(A, B, h) = 0, \quad (5.2.16)$$

$$\lambda \frac{dB}{ds} - N_1(A, B, h) = 0. \quad (5.2.17)$$

Substitution of $A(s)$ and $B(s)$ into Eq. (5.2.12) provides a uniformly valid solution to Eq. (5.2.1), up to terms of order ϵ.

5.3 Conservative Oscillators

The periodic solutions of conservative oscillators have the property that the amplitude of the oscillations are constant [1, 2, 3, 4]. In this section, we use this property as the characteristic defining a conservative oscillator. Our interest is in applying this criterion to the solutions of various discrete models of conservative oscillators. Without loss of generality, we only consider the Duffing equation [15, 16]

$$\frac{d^2y}{dt^2} + y + \epsilon y^3 = 0, \quad 0 < \epsilon \ll 1. \quad (5.3.1)$$

For $\epsilon > 0$, it can be shown that all the solutions to the Duffing differential equation are bounded and periodic [1, 2]. For small ϵ, a uniformly valid expression for $y(t)$ can be calculated by using a variety of perturbation procedures [3, 5]. They all give, to terms of order ϵ, the result

$$y(t) = \bar{a} \sin\left[\left(1 + \frac{3\bar{a}^2\epsilon}{8}\right)t + \phi\right], \quad (5.3.2)$$

where $\bar{a}$ is the constant amplitude and ϕ the constant phase.

We now construct and analyze four discrete models of the Duffing equation for the parameter domain where $0 < \epsilon \ll 1$. Thus, the discrete-multi-time perturbation method of the previous section can be applied. Any finite-difference scheme that has solutions for which the amplitude is not constant will be considered to be an inappropriate discrete model [16].

The four finite-difference models of Eq. (5.3.1) to be investigated are

$$\Gamma y_k + \epsilon y_k^3 = 0, \tag{5.3.3}$$

$$\Gamma y_k + \epsilon y_k^2 \left(\frac{y_{k+1} + y_{k-1}}{2} \right) = 0, \tag{5.3.4}$$

$$\Gamma y_k + \epsilon y_{k-1}^3 = 0, \tag{5.3.5}$$

$$\Gamma y_k + \epsilon y_{k+1}^3 = 0, \tag{5.3.6}$$

where the operator Γ is defined by Eq. (5.2.3). Note that for $\epsilon = 0$, the linear finite-difference scheme obtained

$$\Gamma y_k = 0, \tag{5.3.7}$$

is an exact discrete model for the harmonic oscillator differential equation

$$\frac{d^2 y}{dt^2} + y = 0. \tag{5.3.8}$$

We now present, in detail, the calculations for the perturbation solution to Eq. (5.3.3). The other three examples are done in exactly the same manner.

Comparison of Eqs. (5.2.1) and (5.3.3) gives

$$f(y_{k+1}, y_k, y_{k-1}) = -y_k^3. \tag{5.3.9}$$

Using the result

$$[y_0(k,s)]^3 = [A\cos(hk) + B\sin(hk)]^3 = \left(\frac{3}{4}\right)A(A^2 + B^2)\cos(hk)$$
$$+ \left(\frac{3}{4}\right)B(A^2 + B^2)\sin(hk) + \left(\frac{3}{4}\right)A(A^2 - 3B^2)\cos(3hk)$$
$$+ \left(\frac{3}{4}\right)B(3A^2 - B^2)\sin(3hk), \tag{5.3.10}$$

we find

$$M_1 = -\left(\frac{3}{4}\right)B(A^2 + B^2), \tag{5.3.11}$$

$$N_1 = -\left(\frac{3}{4}\right)A(A^2 + B^2), \tag{5.3.12}$$

and

$$\lambda\frac{dA}{ds} = \left(\frac{3}{4}\right)B(A^2 + B^2), \tag{5.3.13}$$

$$\lambda\frac{dB}{ds} = -\left(\frac{3}{4}\right)A(A^2 + B^2), \tag{5.3.14}$$

where λ is defined in Eq. (5.2.14). Multiplying Eqs. (5.3.13) and (5.3.14), respectively, by A and B, and adding the resulting expressions gives

$$\frac{d}{ds}(A^2 + B^2) = 0, \tag{5.3.15}$$

or

$$\bar{a}^2 = A^2 + B^2 = \text{constant}. \tag{5.3.16}$$

Now define ω to be

$$\omega = \frac{3\bar{a}^2}{4\lambda}. \tag{5.3.17}$$

Then Eqs. (5.3.13) and (5.3.14) become

$$\frac{dA}{ds} = \omega B, \qquad \frac{dB}{ds} = -\omega A. \tag{5.3.18}$$

They have solutions

$$A(s) = \bar{a}\sin(\omega s + \phi), \tag{5.3.19a}$$

$$B(s) = \bar{a}\cos(\omega s + \phi), \tag{5.3.19b}$$

where ϕ is an arbitrary constant. If these results are substituted into Eq. (5.2.12), then the following is obtained

$$y_0(k, s) = \bar{a}\sin(\omega s + hk + \phi). \tag{5.3.20}$$

Now using $s = \epsilon k$, we finally obtain

$$y_k = y_0(k, s) + O(\epsilon) = \bar{a}\sin\left\{1 + \left(\frac{3\bar{a}^2\epsilon}{2}\right)\left[\frac{\sin^2(h/2)}{h\sin(h)}\right]t_k + \phi\right\}, \tag{5.3.21}$$

where $t_k = hk$. Note that in the limits

$$h \to 0, \qquad k \to \infty, \qquad hk = t = \text{fixed}, \tag{5.3.22}$$

the right-side of Eq. (5.3.21) tends to the function of Eq. (5.3.2) as expected.

The significant point is that to first-order in ϵ, the discrete model of the Duffing equation, given by Eq. (5.3.3), has oscillatory solutions with constant amplitude. Thus, using only this criterion, the finite-difference scheme of Eq. (5.3.3) is an adequate model. Note that Eq. (5.3.3) uses a standard local expression for the nonlinear term y^3, i.e.,

$$y^3 \to y_k^3. \tag{5.3.23}$$

However, the linear part of the differential equation is modeled by its exact finite-difference scheme.

Now let $\bar{a}_i$, where $i = (1, 2, 3, 4)$, be the amplitudes of the solutions, respectively, of Eqs. (5.3.3) to (5.3.6). We have just obtained $\bar{a}_1$. Repeating the calculation for the other three cases gives

$$\frac{d}{ds}(\bar{a}_1)^2 = 0, \tag{5.3.24}$$

$$\frac{d}{ds}(\bar{a}_2)^2 = 0, \tag{5.3.25}$$

$$\frac{d}{ds}(\bar{a}_3)^2 = \alpha(\bar{a}_3)^4, \tag{5.3.26}$$

$$\frac{d}{ds}(\bar{a}_4)^2 = -\alpha(\bar{a}_4)^4, \qquad (5.3.27)$$

where

$$\alpha = 3\sin^2\left(\frac{h}{2}\right). \qquad (5.3.28)$$

Observe that the finite-difference schemes of Eqs. (5.3.3) and (5.3.4) give oscillatory solutions for which the amplitude is constant. However, the discrete models of Eqs. (5.3.5) and (5.3.6) have oscillatory solutions whose amplitudes, respectively, increase and decrease, a behavior not consistent with the known properties of the Duffing differential equation. Therefore, we conclude that these discrete models are not appropriate for calculating numerical solutions.

Further examination of Eqs. (5.3.3) and (5.3.4) shows that they have the special symmetry

$$y_{k+1} \leftrightarrow y_{k-1}. \qquad (5.3.29)$$

This result can be generalized in the following manner. Let

$$\Gamma y_k + F(y_{k+1}, y_k, y_{k-1}) = 0, \qquad (5.3.30)$$

be a discrete model of the conservative oscillator differential equation

$$\frac{d^2 y}{dt^2} + f(y) = 0; \qquad (5.3.31)$$

then Eq. (5.3.30) will be an appropriate discrete model if the function $F(y_{k+1}, y_k, y_{k+1})$ has the property

$$F(y_{k+1}, y_k, y_{k-1}) = F(y_{k-1}, y_k, y_{k+1}). \qquad (5.3.32)$$

The existence of this symmetry can be easily understood in terms of the time-reversal invariance of the conservative differential equation. The linear part of the discrete model, Γy_k, automatically satisfies this condition.

There is also a second way of investigating the conservative nature of finite-difference models of differential equations, namely, the use of Taylor series expansions in h to determine the governing differential equation for fixed, but, nonzero values of the step-size. We now demonstrate how this can be accomplished.

In the limits, given by Eq. (5.3.22), all of the above discrete models of the Duffing equation converge [17] to the differential equation. However, in practical calculations, i.e., the numerical integration of the Duffing differential equation, h may be small, but is nevertheless always finite. For this situation, the discrete model corresponds to an entirely different differential equation. To demonstrate this, we make use of the following results:

$$y_k = y(t) + O(h^2), \tag{5.3.33a}$$

$$y_{k\pm1} = y(t) \pm h\frac{dy}{dt} + O(h^2), \tag{5.3.33b}$$

$$\frac{y_{k+1} - 2y_k + y_{k-1}}{4\sin^2\left(\frac{h}{2}\right)} = \frac{d^2y}{dt^2} + O(h^2), \tag{5.3.33c}$$

$$(y_{k\pm1})^3 = y^3 \pm 3hy^2\left(\frac{dy}{dt}\right) + O(h^2). \tag{5.3.33d}$$

Substituting Eqs. (5.3.33) into Eqs. (5.3.3) to (5.3.6) gives, respectively,

$$\frac{d^2y}{dt^2} + y + \epsilon y^3 = O(h^2), \tag{5.3.34}$$

$$\frac{d^2y}{dt^2} + y + \epsilon y^3 = O(h^2), \tag{5.3.35}$$

$$\frac{d^2y}{dt^2} - 3\epsilon hy^2\left(\frac{dy}{dt}\right) + y + \epsilon y^3 = O(h^2), \tag{5.3.36}$$

$$\frac{d^2y}{dt^2} + 3\epsilon hy^2\left(\frac{dy}{dt}\right) + y + \epsilon y^3 = O(h^2). \tag{5.3.37}$$

Examination of these equations shows that to terms of order h^2, the discrete models given by Eqs. (5.3.3) and (5.3.4) represent the conservative Duffing oscillator differential equation. However, the discrete models of Eqs. (5.3.5) and (5.3.6) correspond, to terms of order h^2, respectively, oscillators with negative and positive

damping. This is exactly the result found above using the discrete-two-time perturbation method. Similar results hold for the linear harmonic oscillator differential
equation; see Section 2.3. The symmetric replacement $y^3 \rightarrow y_{k+1}y_k y_{k-1}$ also gives
Eqs. (5.3.34) and (5.3.35).

Consider again the generalized conservative oscillator differential equation

$$\frac{d^2y}{dt^2} + g[y^2, (\dot{y})^2]y = 0, \tag{5.3.38}$$

where $\dot{y} \equiv dy/dt$. In practical applications, the function $g[y^2, (\dot{y})^2]$ is either a
polynomial function or a rational function whose denominator has no real zeros.
A conservative discrete model for Eq. (5.3.38) can be obtained by replacing the
second-derivative with

$$\frac{d^2y}{dt^2} \rightarrow \frac{y_{k+1} - 2y_k + y_{k-1}}{\psi(h)}, \tag{5.3.39a}$$

where

$$\psi(h) = h^2 + O(h^4), \tag{5.3.39b}$$

and replacing the nonlinear function g by a function

$$g[y^2, (\dot{y})^2]y \rightarrow G(y_{k+1}, y_k, y_{k-1}) \tag{5.3.40}$$

having the following two properties

$$G(y_{k+1}, y_k, y_{k-1}) = G(y_{k+1}, y_k, y_{k+1}), \tag{5.3.41a}$$

$$G(y_{k+1}, y_k, y_{k-1}) = g[y^2, (\dot{y})^2]y + O(h^2). \tag{5.3.41b}$$

A particular way to implement this is to replace in $g[y^2, (\dot{y})^2]y$, the y^2 and $(\dot{y})^2$
terms, respectively, by some linear combinations of expressions such as

$$y^2 \rightarrow \begin{cases} y_k^2, \\ \frac{y_{k+1}^2 + y_k^2 + y_{k-1}^2}{3}, \\ \frac{y_{k+1}(y_{k+1} + y_{k-1})}{2}, \\ y_{k+1}y_{k-1}, \\ \text{etc.}, \end{cases} \tag{5.3.42}$$

$$(\dot{y})^2 \rightarrow \begin{cases} \left[\frac{y_{k+1}-y_{k-1}}{2\psi(h)}\right]^2, \\ \left(\frac{1}{2}\right)\left[\frac{y_{k+1}-y_k}{\psi(h)}\right]^2 + \left(\frac{1}{2}\right)\left[\frac{y_k-y_{k-1}}{\psi(h)}\right]^2, \\ \text{etc.,} \end{cases} \tag{5.3.43}$$

where

$$\psi(h) = h + O(h^2). \tag{5.3.44}$$

Classical mechanics defines a conservative oscillator as one for which there exists a constant energy first-integral and all bounded solutions are periodic [1, 2]. Earlier, in Section 3.5, we discussed a best finite-difference scheme for the Duffing oscillator. There our interest was in constructing a discrete model that had a constant (discrete) first integral. We showed that this was possible for the scheme

$$\frac{y_{k+1} - 2y_k + y_{k-1}}{\psi} + y_k + \epsilon(y_k)^2\left(\frac{y_{k+1} + y_{k-1}}{2}\right) = 0, \tag{5.3.45}$$

where

$$\psi = h^2 + O(h^4), \tag{5.3.46}$$

but, not possible for

$$\frac{y_{k+1} - 2y_k + y_{k-1}}{\psi} + y_k + \epsilon y_k^3 = 0. \tag{5.3.47}$$

In this section, an alternative definition of a conservative oscillator was introduced, namely, a second-order differential equation that is invariant under the transformation $t \leftrightarrow -t$. This invariance translates to the requirement that the discrete model should be unchanged when $y_{k+1} \leftrightarrow y_{k-1}$ with the consequence that the amplitude of oscillatory solutions be constant. With this alternative definition, both of the discrete models given by Eqs. (5.3.45) and (5.3.47) are conservative. From the view point of the actual physical phenomena, only the scheme of Eq. (5.3.45) can be used, since the oscillator does in fact satisfy an energy conservation law. This scheme is the one that comes from the application of the nonstandard modeling rules given in Section 3.4.

Finally, for comparison with the first-order in ϵ solution of Eq. (5.3.3), given by Eq. (5.3.21), we give the solution of Eq. (5.3.4). It is

$$y_k = y_0(k, s) + O(\epsilon) = \bar{a} \sin[(1 + \epsilon\alpha)t_k + \phi] + O(\epsilon), \qquad (5.3.48)$$

where

$$\alpha = \left(\frac{3\bar{a}^2}{2}\right)\left[\frac{\cos(h)\sin^2(h/2)}{h\sin(h)}\right], \qquad (5.3.49)$$

and $t_k = hk$.

5.4 Limit-Cycle Oscillators

The nonlinear differential equation of van der Pol [2, 3, 5] serves as an important model equation for one-dimensional dynamic systems having a single, stable limit-cycle. After constructing four finite-difference models, we will use a discrete multi-time perturbation procedure to calculate solutions to the finite-difference equations. A detailed comparison will then be made between these solutions and the corresponding solution of the van der Pol differential equation [18]. One of the issues to be considered is what discrete form should be used to model the first-derivative [19]?

The van der Pol oscillator differential equation is

$$\frac{d^2y}{dt^2} + y = \epsilon(1 - y^2)\frac{dy}{dt}, \qquad \epsilon > 0. \qquad (5.4.1)$$

For the purposes of the present section, we assume that ϵ satisfies the condition

$$0 < \epsilon \ll 1. \qquad (5.4.2)$$

Under this restriction, the limit-cycle is given by the expression [3]

$$y(\theta) = 2\cos\theta + \left(\frac{\epsilon}{4}\right)(3\sin\theta - \sin 3\theta)$$
$$+ \left(\frac{\epsilon^2}{96}\right)(-13\cos\theta + 18\cos 3\theta - 5\cos 5\theta) + O(\epsilon^3), \qquad (5.4.3)$$

where

$$\theta = \left[1 - \frac{\epsilon^2}{16} + O(\epsilon^3)\right] t. \tag{5.4.4}$$

Near the limit-cycle, the solution is [3]

$$y(t) = \frac{2a_0 \cos(t + \phi_0)}{[a_0^2 - (a_0^2 - 4)\exp(-\epsilon t)]^{1/2}} + O(\epsilon), \tag{5.4.5}$$

where a_0 and ϕ are constants. Note that

$$\lim_{t \to \infty} y(t) = 2\cos t + O(\epsilon). \tag{5.4.6}$$

The four discrete models to be investigated are

Model I-A:

$$\Gamma y_k = \epsilon(1 - y_k^2)\left[\frac{y_{k+1} - y_{k-1}}{2h}\right]; \tag{5.4.7}$$

Model I-B:

$$\Gamma y_k = \epsilon(1 - y_k^2)\left[\frac{y_{k+1} - y_{k-1}}{2\left(\frac{4\sin^2(h/2)}{h}\right)}\right]; \tag{5.4.8}$$

Model II:

$$\Gamma y_k = \epsilon\left\{1 - y_k\left[\frac{y_{k+1} + y_{k-1}}{2\cos(h)}\right]\right\}\left[\frac{y_{k+1} - y_{k-1}}{2\sin(h)}\right]; \tag{5.4.9}$$

Model III:

$$\Gamma y_k = \epsilon(1 - y_k^2)\left[\frac{y_k - y_{k-1}}{\left(\frac{4\sin^2(h/2)}{h}\right)}\right]; \tag{5.4.10}$$

where the operator Γ is defined by Eq. (5.2.3). Note that Models I-A, I-B and II use a central difference for the first-derivative, while Model III expresses the first-derivative by a backward Euler. Also, the denominator functions for all but Model I-A are nonstandard. Model II has been studied previously by Potts [20].

Application of the discrete multi-time perturbation method to Eqs. (5.4.7), (5.4.8), (5.4.9) and (5.4.10) gives the following results:

Model I-A:

$$y_k = \frac{2a_0 \cos(t_k + \phi_0)}{[a_0^2 - (a_0^2 - 4)\exp(-\lambda_1 \epsilon t_k)]^{1/2}} + O(\epsilon), \tag{5.4.11a}$$

$$\lambda_1 = \frac{4\sin^2(h/2)}{h^2}; \tag{5.4.11b}$$

Model I-B:

$$y_k = \frac{2a_0\cos(t_k + \phi_0)}{[a_0^2 - (a_0^2 - 4)\exp(-\epsilon t_k)]^{1/2}} + O(\epsilon); \tag{5.4.12}$$

Model II:

$$y_k = \frac{2a_0\cos(t_k + \phi_0)}{[a_0^2 - (a_0^2 - 4)\exp(-\lambda_2\epsilon t_k)]^{1/2}} + O(\epsilon), \tag{5.4.13a}$$

$$\lambda_2 = \frac{4\sin^2(h/2)}{h^2}; \tag{5.4.13b}$$

Model III:

$$y_k = \frac{2a_0\cos(t_k + \phi_0)}{[a_0^2 - (a_0^2 - 4)\exp(-\epsilon t_k)]^{1/2}} + O(\epsilon). \tag{5.4.14}$$

To illustrate how the above results were obtained, we show the details of the calculation for Model I-A. From Section 5.2, the function $f(y_{k+1}, y_k, y_{k-1})$ is

$$f = y_k^2\left[\frac{y_{k+1} - y_{k-1}}{2h}\right]. \tag{5.4.15}$$

From this, the amplitude functions $A(s)$ and $B(s)$ are determined by the equations

$$\frac{dA}{ds} = -\left[\frac{\sin^2(h/2)}{2h}\right]A(A^2 + B^2 - 4), \tag{5.4.16}$$

$$\frac{dB}{ds} = -\left[\frac{\sin^2(h/2)}{2h}\right]B(A^2 + B^2 - 4). \tag{5.4.17}$$

Define z as

$$z = A^2 + B^2,$$

multiply Eqs. (5.4.16) and (5.4.17), respectively, by A and B, and add the resulting expressions to obtain the following result

$$\frac{dz}{ds} = -\frac{\sin^2(h/2)}{h}z(z - 4). \tag{5.4.18}$$

Let $z(0) = z_0$, then Eq. (5.4.18) has the solution

$$z(s) = \frac{4z_0}{z_0 - (z_0 - 4)\exp(-\bar{\lambda}_1 s)}, \tag{5.4.19}$$

$$\bar{\lambda}_1 = \frac{4\sin^2(h/2)}{h}. \tag{5.4.20}$$

Since $s = \epsilon k$ and $t_k = hk$, then

$$\bar{\lambda}_1 s = \lambda_1 \epsilon t_k, \tag{5.4.21}$$

where

$$\lambda_1 = \frac{4\sin^2(h/2)}{h^2}. \tag{5.4.22}$$

Now Eq. (5.2.12) can be rewritten to the form

$$y_0 = (k, s) = A(s)\cos(t_k) + B(s)\sin(t_k) = a(s)\cos[t_k + \phi(s)], \tag{5.4.23}$$

where

$$a^2 = A^2 + B^2, \qquad \tan\phi = -\frac{B}{A}. \tag{5.4.24}$$

From Eqs. (5.4.16) and (5.4.17), it follows that

$$\frac{dA}{dB} = \frac{A}{B}, \tag{5.4.25}$$

or

$$A(s) = cB(s), \tag{5.4.26}$$

where c is an arbitrary constant. This shows that the phase $\phi(s)$ is a constant, i.e.,

$$\phi(s) = \phi_0 = \text{constant}. \tag{5.4.27}$$

Putting all these results together gives Eqs. (5.4.11).

Examination of the four discrete models for the van der Pol equation indicates that under the condition of Eq. (5.4.2), Models I-B and III give the same result as the perturbation solution to the van der Pol differential equation. However, for all four models, we have the proper convergence, i.e.,

$$\lim_{\substack{h \to 0 \\ k \to \infty \\ hk=t=\text{fixed}}} y_k = y(t). \tag{5.4.28}$$

136

Note that both Models I-B and III use nonstandard denominator functions for the discrete first-derivative.

The question now arises: Can a principle or requirement be formulated that will allow us to select either Model I-B or Model III? The answer to this question is yes. The requirement will be that the lowest order term in the Taylor series expansion (in h) of the discrete model should reproduce the original van der Pol differential equation. We used this technique in Section 5.3 to explain the behavior of various discrete models of the Duffing equation.

Let $\psi(h)$ be defined as

$$\psi(h) \equiv \frac{4\sin^2(h/2)}{h}.$$ (5.4.29)

An elementary calculation gives

$$\Gamma y_k = \frac{d^2y}{dt} + y + O(h^2),$$ (5.4.30)

$$1 - y_k^2 = 1 - y^2 + O(h^2),$$ (5.4.31)

$$\frac{y_{k+1} - y_{k-1}}{2\psi(h)} = \frac{dy}{dt} + O(h^2),$$ (5.4.32)

$$\frac{y_k - y_{k-1}}{\psi(h)} = \frac{dy}{dt} - \left(\frac{h}{2}\right)\frac{d^2y}{dt^2} + O(h^2).$$ (5.4.33)

From these expressions, it follows that:

Model I-B:

$$\Gamma y_k - \epsilon(1 - y_k^2)\left[\frac{y_{k+1} - y_{k-1}}{2\psi(h)}\right] = \left[\frac{d^2y}{dt^2} + y - \epsilon(1 - y^2)\frac{dy}{dt}\right] + O(h^2);$$ (5.4.34)

Model III:

$$\Gamma y_k - \epsilon(1 - y_k^2)\left[\frac{y_k - y_{k-1}}{\psi(h)}\right]$$
$$= \left[\frac{d^2y}{dt^2} + y - \epsilon(1 - y^2)\frac{dy}{dt} + \left(\frac{\epsilon h}{2}\right)(1 - y^2)\frac{d^2y}{dt^2}\right] + O(h^2).$$ (5.4.35)

Note that to $O(h^2)$, Model I-B gives the van der Pol differential equation. However, this does not occur for Model III since there is an $O(h)$ term. Thus, the finite-difference scheme of Model III describes to $O(h^2)$ the modified van der Pol equation

$$\frac{d^2y}{dt^2} + y = \epsilon(1-y^2)\frac{dy}{dt} - \left(\frac{\epsilon h}{2}\right)(1-y^2)\frac{d^2y}{dt^2}.\tag{5.4.36}$$

Using the method of harmonic balance [21], the influence of the extra term on the right-side of Eq. (5.4.36) can be determined. The approximate solution to the limit-cycle of the van der Pol equation is [21]

$$y_{\text{VDP}}(t) = 2\cos t.\tag{5.4.37}$$

The same calculation applied to the above modified van der Pol equation gives

$$y_{\text{MVDP}}(t) = 2\cos\left[\left(1 + \frac{\epsilon h}{2}\right)t\right].\tag{5.4.38}$$

Note that besides a small shift in the frequency, the modified van der Pol equation has essentially the same properties as the usual van der Pol oscillator. However, because of the extra term that Model III introduces, our preference would be to select Model I-B as the finite difference scheme to use in the numerical integration of the van der Pol differential equation.

5.5 General Oscillator Equations

A large class of one-dimensional, nonlinear oscillators can be modeled by differential equations having the form [1–6]

$$\frac{d^2y}{dt^2} + y + f(y^2)\frac{dy}{dt} + g(y^2)y = 0.\tag{5.5.1}$$

For example, the Duffing equation corresponds to

$$f(y^2) = 0, \qquad g(y^2) = \epsilon y^2,\tag{5.5.2}$$

138

while the van der Pol equation has

$$f(y^2) = -\epsilon(1 - y^2), \qquad g(y^2) = 0. \tag{5.5.3}$$

The results of the previous two sections suggest the following nonstandard finite-difference scheme for Eq. (5.5.1)

$$\frac{y_{k+1} - 2y_k + y_{k-1}}{4\sin^2\left(\frac{h}{2}\right)} + y_k + f(y_k^2)\left[\frac{y_{k+1} - y_{k-1}}{\frac{4\sin^2\left(\frac{h}{2}\right)}{h}}\right] + g(y_k^2)\left(\frac{y_{k+1} + y_{k-1}}{2}\right) = 0. \tag{5.5.4}$$

Note that Eq. (5.5.4) has the features:

(i) When $f(y^2) = 0$ and $g(y^2) = 0$, the discrete model is an exact finite-difference scheme for the harmonic oscillator equation

$$\frac{d^2y}{dt^2} + y = 0. \tag{5.5.5}$$

(ii) If $f(y^2) = 0$, then Eq. (5.5.1) is the equation of motion for a conservative oscillator. The discrete form for $g(y^2)y$ is constructed on the results obtained in Section 5.3 for conservative oscillators.

(iii) The form for $g(y^2)y = 0$ is consistent with the analysis of the van der Pol oscillator as discussed in Section 5.4.

(iv) Of importance is that the discrete model given by Eq. (5.5.4) is explicit, i.e., y_{k+2} can be solved for and expressed in terms of y_k and y_{k+1}. Consequently, a numerical evaluation of Eq. (5.5.4) involves only a two-step iteration procedure, i.e., shifting the index by one and solving Eq. (5.5.4) for y_{k+2} gives

$$y_{k+2} = \frac{[2\cos(h)]y_{k+1} - \left[1 - hf(y_{k+1}^2) + 4\sin^2\left(\frac{h}{2}\right) \cdot g(y_{k+1}^2)\right]y_k}{1 + hf(y_{k+1}^2) + 4\sin^2\left(\frac{h}{2}\right) \cdot g(y_{k+1}^2)}. \tag{5.5.6}$$

5.6 Response of a Linear System [22]

The linear, damped oscillator can be used to model a large number of physical systems in the physical [1, 4] and engineering sciences [23, 24]. In particular, such a class of systems arises in civil earthquake engineering [24–26]. For a one-dimensional system, the equation of motion takes the form

$$\frac{d^2y}{dt^2} + 2c\omega\frac{dy}{dt} + \omega^2 y = g(t), \tag{5.6.1}$$

where ω is the angular frequency of the undamped system, c is related to the damping coefficient, and $g(t)$ is a forcing function. In many applications, the forcing function is measured or known only at fixed, equal time intervals, i.e.,

$$g(t) \rightarrow g_k, \qquad t_k = hk, \tag{5.6.2}$$

where k is an integer and $h = \Delta t$ is the interval between measurement of $g(t)$. Thus, with this limitation in mind, a discrete form of the left-side is required to have Eq. (5.6.1) make physical sense. This section gives a general computational procedure for calculating "$y(t)$" [22] and generalizes the work presented by Ly [27].

To proceed, we rewrite Eq. (5.6.1) in the dimensionless form [3]

$$\frac{d^2y}{dt^2} + 2\epsilon\frac{dy}{dt} + y = f(t), \tag{5.6.3}$$

where ϵ is the dimensionless damping constant and t is a dimensionless time variable. Selection of the initial conditions

$$y(t_0) = y_0, \qquad \dot{y}(t_0) = \dot{y}_0, \tag{5.6.4}$$

where the "dots" indicate time derivatives, we can express the general solution to Eqs. (5.6.3) and (5.6.4) as [28]

$$\begin{aligned}
y(t) = {}&M(t)[(\beta\cos\beta t_0 - \epsilon\sin\beta t_0)y_0 - (\sin\beta t_0)\dot{y}_0]\cos\beta t \\
&+ M(t)[(\epsilon\cos\beta t_0 + \beta\sin\beta t_0)y_0 + (\cos\beta t_0)\dot{y}_0]\sin\beta t \\
&+ N(t)\int_{t_0}^{t} f(s)e^{\epsilon s}[(\sin\beta t)\cos\beta s - (\cos\beta t)\sin\beta s]ds,
\end{aligned} \tag{5.6.5}$$

where

$$\begin{cases} M(t) = \exp[-\epsilon(t - t_0)]/\beta, \\ N(t) = \exp(-\epsilon t)/\beta, \\ \beta^2 = 1 - \epsilon^2. \end{cases} \quad (5.6.6)$$

The derivative, $\dot{y}(t)$, can be found by differentiating Eq. (5.6.5) using the results of Eq. (5.6.6).

If $y_1(t)$ and $y_2(t)$ are defined to be

$$y_1(t) \equiv y(t), \qquad y_2(t) \equiv \dot{y}(t), \quad (5.6.7)$$

then Eq. (5.6.3) takes the form

$$\frac{dy_1}{dt} = y_2, \quad (5.6.8)$$

$$\frac{dy_2}{dt} = -2\epsilon y_2 - y_1 + f(t). \quad (5.6.9)$$

The exact finite-difference scheme [29] for this system of first-order differential equations can be determined in a straightforward, but, long calculation. It is [22]

$$\begin{aligned} y_1(k+1) = {} & R[\beta \cos \beta h + \epsilon \sin \beta h]y_1(k) + R[\sin \beta h]y_2(k) \\ & + S[\sin \beta h(k+1)] \int_{hk}^{h(k+1)} f(s)e^{\epsilon s} \cos \beta s\, ds \\ & - S[\cos \beta h(k+1)] \int_{hk}^{h(k+1)} f(s)e^{\epsilon s} \sin \beta s\, ds, \quad (5.6.10) \end{aligned}$$

$$\begin{aligned} y_2(k+1) = {} & R[\sin \beta h]y_1(k) + R[-\epsilon \sin \beta h + \beta \cos \beta h]y_2(k) \\ & + S[\beta \cos \beta h(k+1) - \epsilon \sin \beta h(k+1)] \int_{hk}^{k(k+1)} f(s)e^{\epsilon s} \sin \beta s\, ds \\ & + S[\epsilon \cos \beta h(k+1) + \beta \sin \beta h(k+1)] \int_{hk}^{h(k+1)} f(s)e^{\epsilon s} \cos \beta s\, ds, \quad (5.6.11) \end{aligned}$$

where

$$\begin{cases} R = \exp(-\epsilon h)/\beta, \\ S = \exp[-\epsilon h(k+1)]/\beta, \end{cases} \quad (5.6.12)$$

and

$$y_1(k) = y_1(t_k), \qquad y_2(k) = y_2(t_k). \tag{5.6.13}$$

Let the forcing function, $f(t)$, be known only at the discrete times $t_k = hk$. What should we do in this situation? The simplest way to proceed is to replace $f(s)$ in the integrands of Eqs. (5.6.10) and (5.6.11) by a linear functional form in each time interval t_k to t_{k+1}. This gives essentially the results obtained by Ly [27]. However, if additional information is available or if a nonlinear variation of f is used in each time interval [30], then Eqs. (5.6.10) and (5.6.11) determine the corresponding discrete model once the integrals are evaluated. Also, the acceleration, $\ddot{y}(t)$, can be calculated by making use of Eq. (5.6.3), i.e.,

$$\ddot{y}(t) \equiv y_3(t) = -2\epsilon\dot{y}(t) - y(t) + f(t). \tag{5.6.14}$$

Therefore, $\ddot{y}(t)$ can be determined from a knowledge of $y_1(k)$ and $y_2(k)$, by using the relation

$$y_3(k) = -2\epsilon y_2(k) - y_1(k) + f_k. \tag{5.6.15}$$

Finally, the finite-difference scheme constructed above should be stable since it was obtained from the exact discrete model of the unforced oscillator.

References

1. H. Goldstein, *Classical Mechanics* (Addison-Wesley; Reading, MA; 1980).

2. J. J. Stoker, *Nonlinear Vibrations* (Interscience, New York, 1987).

3. R. E. Mickens, *Nonlinear Oscillations* (Cambridge University Press, New York, 1981).

4. A. B. Pippard, *The Physics of Vibration I* (Cambridge University Press, Cambridge, 1978).

5. A. H. Nayfeh, *Problems in Perturbation* (Wiley-Interscience, New York, 1985).

6. N. Bessis and G. Bessis, *Journal of Mathematical Physics* **21**, 2780-2791 (1980). A note on the Schrödinger equation for the $x^2 + \lambda x^2/(1 + gx^2)$ potential.

7. R. E. Mickens, *Journal of the Franklin Institute* **321**, 39–47 (1986). Periodic solutions of second-order nonlinear difference equations containing a small parameter III. Perturbation theory.

8. R. E. Mickens, *Journal of Franklin Institute* **324**, 263–271 (1987). Periodic solutions of second-order nonlinear difference equations containing a small parameter IV. Multi-discrete time method.

9. D. Greenspan, *Discrete Models* (Addison-Wesley; Reading, MA; 1973).

10. D. Greenspan, *Arithmetic Applied Mathematics* (Pergamon, New York, 1980).

11. Y. Shibbern, *Discrete-Time Hamiltonian Dynamics* (Ph.D. Thesis, University of Texas at Arlington, 1992).

12. R. Labudde, *International Journal of General Systems* **6**, 3–12 (1980). Discrete Hamiltonian mechanics.

13. T. D. Lee, *Journal of Statistical Physics* **46**, 843–860 (1987). Difference equations and conservation laws.

14. Y. Wu, *Computers and Mathematics with Applications* **20**, 61–75 (1990). The discrete variational approach to the Euler-Lagrange equations.

15. R. E. Mickens, O. Oyedeji and C. R. McIntyre, *Journal of Sound and Vibration* **130**, 509–512 (1989). A difference equation model of the Duffing equation.

16. R. E. Mickens, *Journal of Sound and Vibration* **124**, 194–198 (1988). Properties of finite difference models of non-linear conservative oscillators.

17. H. J. Stetter, *Analysis of Discretization Methods for Ordinary Differential Equations* (Springer-Verlag, Berlin, 1973).

18. R. E. Mickens, Investigation of finite-difference models of the van der Pol equation, in *Differential Equations and Applications*, A. R. Aftabizadeh, editor (Ohio University Press; Columbus, OH; 1988), pp. 210–215.

19. R. E. Mickens and A. Smith, *Journal of the Franklin Institute* **327**, 143–149 (1990). Finite-difference models of ordinary differential equations: Influence of denominator functions.

20. R. B. Potts, *Nonlinear Analysis* **7**, 801–812 (1983). Van der Pol difference equation.

21. R. E. Mickens, *Journal of Sound and Vibration* **94**, 456–460 (1984). Comments on the method of harmonic balance.

22. R. E. Mickens, *Journal of Sound and Vibration* **112**, 183–186 (1987). A computational method for the determination of the response of a linear system.

23. L. A. Pipes and L. R. Harvill, *Applied Mathematics for Engineers and Physicists* (McGraw-Hill, New York, 1970).

24. R. W. Clough and J. Penzien, *Dynamics of Structure* (McGraw-Hill, New York, 1978).

25. N. C. Nigam and P. C. Jennings, *Bulletin of the Seismological Society of America* **59**, 909–922 (1969). Calculation of response spectra from strong earthquake records.

26. M. P. Singh and M. Ghafory-Ashtiany, *Earthquake Engineering and Structural Dynamics* **14**, 133–146 (1986). Modal time history analysis of non-classically damped structures for seismic motions.

27. B. L. Ly, *Journal of Sound and Vibration* **95**, 435–438 (1984). A computation technique for the response of linear systems.

28. E. A. Kraut, *Fundamentals of Mathematical Physics* (McGraw-Hill, New York, 1967).

29. R. E. Mickens, Difference equation models of differential equations having zero local truncation errors, in *Differential Equations*, I. W. Knowles and R. T. Lewis, editors (North-Holland, Amsterdam, 1984).

30. T. R. F. Nonweiler, *Computational Mathematics: An Introduction to Numerical Approximation* (Halsted Press, New York, 1984).

Chapter 6

TWO FIRST-ORDER, COUPLED ORDINARY

DIFFERENTIAL EQUATIONS

6.1 Introduction

In this chapter, we construct a class of finite-difference schemes for two coupled first-order ordinary differential equations. These schemes have linear stability properties that are the same as the differential equation for all step-sizes. The differential equations considered are assumed to have a single (real) fixed-point. A major consequence of these schemes is the absence of elementary numerical instabilities. Briefly, numerical instabilities are solutions to the discrete finite-difference equations that do not correspond to any of the solutions of the original differential equations [1, 2, 3]. (See Chapter 2.) Thus, the given finite-difference equations are not able to model the correct mathematical properties of the solutions to the differential equations [2]. In general, numerical instabilities will occur when the linear stability properties of the corresponding fixed-points of the differential and difference equations do not agree [4].

The work in this chapter extends the results of Mickens and Smith [3] and Mickens [4] to the case of the two coupled, first-order ordinary differential equations

$$\frac{dx}{dt} = F(x, y), \tag{6.1.1}$$

$$\frac{dy}{dt} = G(x, y), \tag{6.1.2}$$

that have only a single (real) fixed-point which can be chosen to be at the origin, i.e., $(\bar{x}, \bar{y}) = (0, 0)$, where $\bar{x}$ and $\bar{y}$ are simultaneous solutions to the equations

$$F(\bar{x}, \bar{y}) = 0 = G(\bar{x}, \bar{y}). \tag{6.1.3}$$

After presenting certain background information in Section 6.2, we demonstrate how to explicitly construct discrete models for the linear parts of Eqs. (6.1.1) and (6.1.2) that have the correct linear stability properties for all values of the step-size. This result is then applied in Section 6.4 to the full nonlinear differential equation where we find that two major classes of discrete models emerge: the fully explicit and semi-explicit schemes. Section 6.5 gives a variety of examples of finite-difference schemes constructed according to the rules of Section 6.4. Further generalizations and modifications of these rules are discussed for individual equations.

Finally, it should be mentioned that while the coupled differential equation systems considered in this chapter are a small subset of all possible such equations, they do describe many important dynamical systems. Examples include the van der Pol limit-cycle oscillator [5, 6]

$$\frac{dx}{dt} = y, \qquad \frac{dy}{dt} = -x + \epsilon(1 - x^2)y, \qquad (6.1.4)$$

where ϵ is a positive parameter; the Lewis oscillator [7, 8]

$$\frac{dx}{dt} = y, \qquad \frac{dy}{dt} - x + \epsilon(1 - |x|)y; \qquad (6.1.5)$$

the Duffing oscillator [6]

$$\frac{dx}{dt} = y, \qquad \frac{dy}{dt} = -x - \lambda x^3, \qquad (6.1.6)$$

where λ is a parameter; and the modeling of batch fermentation processes [9]

$$\frac{dx}{dt} = -(A\alpha\beta)y - [(A\alpha)xy + (A\beta)y^2] - Axy^2, \qquad (6.1.7)$$

$$\frac{dy}{dt} = [(B\alpha\beta)x + (B\gamma\beta^2)y] + [(\alpha B)x^2 + (2B\beta\gamma)xy - (B\beta^2\epsilon)y^2]$$
$$+ [(B\gamma)x^2y - (2B\beta\epsilon)xy^2] - (B\epsilon)x^2y^2. \qquad (6.1.8)$$

Other systems that can be modeled by two coupled, first-order ordinary differential equations arise in the biological sciences [10, 11], chemistry [12], and engineering [13, 14].

6.2 Background

In more detail, we assume that Eqs. (6.1.1) and (6.1.2) take the form

$$\frac{dx}{dt} = ax + by + f(x,y) \equiv F(x,y), \tag{6.2.1}$$

$$\frac{dy}{dt} = cx + dy + g(x,y) \equiv G(x,y), \tag{6.2.2}$$

where

$$ad - bc \neq 0, \tag{6.2.3}$$

and

$$f(x,y) = O(x^2 + y^2), \qquad g(x,y) = O(x^2 + y^2). \tag{6.2.4}$$

We further assume that the only (real) solution to the equations

$$F(\bar{x},\bar{y}) = 0, \qquad G(\bar{x},\bar{y}) = 0, \tag{6.2.5}$$

is

$$\bar{x} = 0, \qquad \bar{y} = 0. \tag{6.2.6}$$

In general, for dynamical systems that model actual physical phenomena, the functions $F(x,y)$ and $G(x,y)$ are real analytic functions of x and y.

The concept of an *exact finite-difference scheme* has already been defined and discussed in Section 3.2. The following theorem will be of value to the calculations given in Section 6.3.

Theorem. Assume that the system of two coupled ordinary differential equations

$$\frac{dX}{dt} = \Gamma(X), \qquad X(t_0) \equiv X_0, \tag{6.2.7}$$

where

$$X = \begin{pmatrix} x \\ y \end{pmatrix}, \qquad \Gamma(X) = \begin{pmatrix} F(x,y) \\ G(x,y) \end{pmatrix}, \tag{6.2.8}$$

has the solution

$$X(t) = \Sigma(X_0, t_0, t). \tag{6.2.9}$$

Then Eq. (6.2.7) has the exact difference scheme

$$X_{k+1} = \Sigma[X_k, hk, h(k+1)], \tag{6.2.10}$$

where

$$X_k = X(hk). \tag{6.2.11}$$

We will now use this theorem "in reverse" to construct the exact finite-difference scheme for the linear parts of Eqs. (6.2.1) and (6.2.2). See also Section 3.3.

6.3 Exact Scheme for Linear Ordinary Differential Equations

The terms in Eqs. (6.2.1) and (6.2.2) correspond to the following differential equation system

$$\frac{du}{dt} = au + bw, \tag{6.3.1}$$

$$\frac{dw}{dt} = cu + dw. \tag{6.3.2}$$

With the initial conditions

$$u_0 = u(t_0), \qquad w_0 = w(t_0), \tag{6.3.3}$$

and the requirement

$$ad - bc \neq 0, \tag{6.3.4}$$

the general solution to Eqs. (6.3.1) and (6.3.2) are given by the relations [15]

$$u(t) = -\left(\frac{b}{\lambda_1 - \lambda_2}\right)\left[\left(\frac{\lambda_2 - a}{b}\right)u_0 - w_0\right]e^{\lambda_1(t-t_0)}$$
$$+ \left(\frac{b}{\lambda_1 - \lambda_2}\right)\left[\left(\frac{\lambda_1 - a}{b}\right)u_0 - w_0\right]e^{\lambda_2(t-t_0)}, \qquad (6.3.5)$$

$$w(t) = -\left(\frac{\lambda_1 - a}{\lambda_1 - \lambda_2}\right)\left[\left(\frac{\lambda_2 - a}{b}\right)u_0 - w_0\right]e^{\lambda_1(t-t_0)}$$
$$+ \left(\frac{\lambda_2 - a}{\lambda_1 - \lambda_2}\right)\left[\left(\frac{\lambda_1 - a}{b}\right)u_0 - w_0\right]e^{\lambda_2(t-t_0)}, \qquad (6.3.6)$$

where

$$2\lambda_{1,2} = (a+d) \pm \sqrt{(a+d)^2 - 4(ad - bc)}. \qquad (6.3.7)$$

The exact difference scheme for Eqs. (6.3.1) and (6.3.2) is obtained by making the following substitutions in Eqs. (6.3.5) and (6.3.6):

$$\begin{cases} t_0 \to t_k = hk, \\ t \to t_{k+1} = h(k+1), \\ u_0 \to u_k, \\ u(t) \to u_{k+1}, \\ w_0 \to w_k, \\ w(t) \to w_{k+1}. \end{cases} \qquad (6.3.8)$$

The results of these substitutions are

$$\frac{u_{k+1} - \psi u_k}{\phi} = au_k + bw_k, \qquad (6.3.9)$$

$$\frac{w_{k+1} - \psi w_k}{\phi} = cu_k + dw_k, \qquad (6.3.10)$$

where

$$\psi \equiv \psi(\lambda_1, \lambda_2, h) = \frac{\lambda_1 e^{\lambda_2 h} - \lambda_2 e^{\lambda_1 h}}{\lambda_1 - \lambda_2}, \qquad (6.3.11)$$

$$\phi \equiv \phi(\lambda_1, \lambda_2, h) = \frac{e^{\lambda_1 h} - e^{\lambda_2 h}}{\lambda_1 - \lambda_2}. \tag{6.3.12}$$

The left-sides of Eqs. (6.3.9) and (6.3.10) are the discrete first-derivatives for Eqs. (6.3.1) and (6.3.2), i.e.,

$$\frac{du}{dt} \rightarrow \frac{u_{k+1} - \psi u_k}{\phi}, \tag{6.3.13}$$

$$\frac{dw}{dt} \rightarrow \frac{w_{k+1} - \psi w_k}{\phi}. \tag{6.3.14}$$

In the limit as

$$h \rightarrow 0, \qquad k \rightarrow \infty, \qquad hk = t = \text{fixed}, \tag{6.3.15}$$

these discrete derivatives reduce to the standard definition of the first-derivative since

$$\psi(\lambda_1, \lambda_2, h) = 1 + O(h^2), \tag{6.3.16}$$

$$\phi(\lambda_1, \lambda_2, h) = h + O(h^2). \tag{6.3.17}$$

However, note that for fixed, finite values of h, the nonstandard discrete derivatives given by Eqs. (6.3.13) and (6.3.14), do not agree with the definition of the discrete first-derivatives

$$\frac{du}{dt} \rightarrow \frac{u_{k+1} - u_k}{h}, \tag{6.3.18}$$

$$\frac{dw}{dt} \rightarrow \frac{w_{k+1} - w_k}{h}, \tag{6.3.19}$$

as given by the standard procedures [1, 3].

In summary, the finite-difference model given by Eqs. (6.3.9) and (6.3.10) is the exact difference equation representation of Eqs. (6.3.1) and (6.3.2). As such, they satisfy, for any step-size h, the conditions

$$u_k = u(hk), \qquad w_k = w(hk), \tag{6.3.20}$$

where $u(t)$ and $w(t)$ are the solutions to Eqs. (6.3.1) and (6.3.2), and u_k and w_k are the solutions to Eqs. (6.3.9) and (6.3.10).

6.4 Nonlinear Equations

The simplest finite-difference scheme for the coupled, first-order nonlinear differential equations given by Eqs. (6.2.1) and (6.2.2) that has the correct linear stability properties for all values of the step-size is

$$\frac{x_{k+1} - \psi x_k}{\phi} = ax_k + by_k + f(x_k, y_k), \tag{6.4.1}$$

$$\frac{y_{k+1} - \psi y_k}{\phi} = cx_k + dy_k + g(x_k, y_k), \tag{6.4.2}$$

where ϕ and ψ are defined by Eqs. (6.3.11) and (6.3.12). This scheme evaluates the functions $f(x, y)$ and $g(x, y)$ at the same computational grid point, i.e., (x_k, y_k). Consequently, the discrete model of Eqs. (6.4.1) and (6.4.2) is explicit, i.e., both x_{k+1} and y_{k+1} are determined directly in terms of x_k and y_k.

A second possibility is the scheme

$$\frac{x_{k+1} - \psi x_k}{\phi} = ax_k + by_k + f(x_k, y_k), \tag{6.4.3}$$

$$\frac{y_{k+1} - \psi y_k}{\phi} = cx_k + dy_k + g(x_{k+1}, y_k). \tag{6.4.4}$$

Comparison with the previous discrete model shows that while Eq. (6.4.3) is the same as Eq. (6.4.1), Eqs. (6.4.4) and (6.4.2) differ. This difference occurs because in the $g(x, y)$ function the "x" variable is replaced by x_k in Eq. (6.4.2), but, by x_{k+1} in Eq. (6.4.4). This second scheme is a semi-explicit discrete model. By this we mean that for given values of (x_k, y_k), the value of x_{k+1} is first calculated from Eq. (6.4.3), then y_{k+1} is determined by Eq. (6.4.4). Thus, there is a definite order to how the calculation should be done.

In general, a variety of other discrete models can exist for Eqs. (6.2.1) and (6.2.2). In generic form, we indicate their structure by

$$\frac{x_{k+1} - \psi x_k}{\phi} = ax_k + by_k + f_k, \tag{6.4.5}$$

$$\frac{y_{k+1} - \psi y_k}{\phi} = cx_k + dy_k + g_k, \qquad (6.4.6)$$

where f_k and g_k denote the particular discrete forms selected for $f(x,y)$ and $g(x,y)$. The important point is that all these schemes, including Eqs. (6.4.1) and (6.4.2), and Eqs. (6.4.3) and (6.4.4), have the property that their fixed-point at $(\bar{x}, \bar{y}) = (0,0)$ has exactly the same linear stability behavior as the differential equation system for all step-sizes. Since the elementary numerical instabilities arise from a change in the linear stability properties of the fixed-points, it follows that these schemes will not have elementary numerical instabilities for any step-size.

In the section to follow, we will use the above results to construct nonstandard finite-difference models for a number of differential equations.

6.5 Examples

In this section discrete models of both conservative and limit-cycle oscillator systems that can be given as two coupled, first-order differential equations will be studied.

6.5.1 Harmonic Oscillator

The second-order harmonic oscillator differential equation is

$$\frac{d^2x}{dt^2} + x = 0. \qquad (6.5.1)$$

In system form it becomes

$$\frac{dx}{dt} = y, \qquad (6.5.2)$$

$$\frac{dy}{dt} = -x. \qquad (6.5.3)$$

Comparing to Eqs. (6.2.1) and (6.2.2) gives

$$a = 0, \qquad b = 1, \qquad c = -1, \qquad d = 0, \qquad (6.5.4a)$$

$$f(x,y) = 0, \qquad g(x,y) = 0. \qquad (6.5.4b)$$

152

Substitution of these results into Eqs. (6.3.11) and (6.3.12) gives

$$\phi = \sin(h), \qquad \psi = \cos(h). \tag{6.5.5}$$

Consequently, the exact difference scheme for the harmonic oscillator is

$$\frac{x_{k+1} - \cos(h)x_k}{\sin(h)} = y_k, \tag{6.5.6}$$

$$\frac{y_{k+1} - \cos(h)y_k}{\sin(h)} = -x_k. \tag{6.5.7}$$

A single second-order difference equation can be obtained by substituting the y_k of Eq. (6.5.6) into Eq. (6.5.7). This gives

$$\frac{[x_{k+2} - \cos(h)x_{k+1}] - \cos(h)[x_{k+1} - \cos(h)x_k]}{\sin^2(h)} = -x_k, \tag{6.5.8}$$

and

$$x_{k+2} - 2\cos(h)x_{k+1} + [\cos^2(h) + \sin^2(h)]x_k = 0. \tag{6.5.9}$$

But,

$$\cos^2(h) + \sin^2(h) = 1, \tag{6.5.10a}$$

$$2\cos(h) = 2 - 4\sin^2\left(\frac{h}{2}\right); \tag{6.5.10b}$$

therefore, Eq. (6.5.9) can be written as

$$(x_{k+1} - 2x_k + x_{k-1}) + \left[4\sin^2\left(\frac{h}{2}\right)\right]x_k = 0, \tag{6.5.11}$$

or, finally

$$\frac{x_{k+1} - 2x_k + x_{k-1}}{4\sin^2\left(\frac{h}{2}\right)} + x_k = 0. \tag{6.5.12}$$

This is precisely the form found earlier in Eq. (3.3.40).

6.5.2 Damped Harmonic Oscillator

The damped linear oscillator is described by the differential equation

$$\frac{d^2x}{dt^2} + 2\epsilon\frac{dx}{dt} + x = 0, \tag{6.5.13}$$

where ϵ is a positive constant. In system form this becomes

$$\frac{dx}{dt} = y, \tag{6.5.14}$$

$$\frac{dy}{dt} = -x - 2\epsilon y. \tag{6.5.15}$$

For this case, we have

$$a = 0, \qquad b = 1, \qquad c = -1, \qquad d = -2\epsilon, \tag{6.5.16}$$

and the ψ and ϕ functions are

$$\psi(\epsilon, h) = \frac{\epsilon e^{-\epsilon h}}{\sqrt{1 - \epsilon^2}} + e^{-\epsilon h} \cos\left(\sqrt{1 - \epsilon^2}\right) h, \tag{6.5.17}$$

$$\phi(\epsilon, h) = \frac{e^{-\epsilon h}}{\sqrt{1 - \epsilon^2}} \cdot \sin\left(\sqrt{1 - \epsilon^2}\right) h. \tag{6.5.18}$$

Thus, the exact finite-difference scheme for the damped linear oscillator is

$$\frac{x_{k+1} - \psi x_k}{\phi} = y_k, \tag{6.5.19}$$

$$\frac{y_{k+1} - \psi y_k}{\phi} = -x_k - 2\epsilon y_k. \tag{6.5.20}$$

Using the expression for y_k, given by Eq. (6.5.19), a single second-order equation can be obtained for x_k; it is

$$\frac{x_{k+1} - 2\psi x_k + \psi^2 x_{k-1}}{\phi^2} + 2\epsilon\left(\frac{x_k - \psi x_{k-1}}{\phi}\right) + x_{k-1} = 0, \tag{6.5.21}$$

where ψ and ϕ are given by Eqs. (6.5.17) and (6.5.18). An alternative form can be determined by transforming the various terms of Eq. (6.5.21). For example, multiplying by ϕ^2

$$[x_{k+1} - 2\psi x_k + \psi^2 x_{k-1}] + 2\epsilon\phi(x_k - \psi x_{k-1}) + \phi^2 x_{k-1} = 0, \tag{6.5.22}$$

154

and using the relations

$$-2\psi x_k = -2x_k + 2(1-\psi)x_k, \tag{6.5.23}$$

$$(\phi^2 + \psi^2)x_{k-1} = x_{k-1} + (\phi^2 + \psi^2 - 1)x_{k-1}, \tag{6.5.24}$$

gives

$$(x_{k+1} - 2x_k + x_{k-1}) + 2\epsilon\phi(x_k - \psi x_{k-1}) + 2(1-\psi)x_k$$
$$+ (\phi^2 + \psi^2 - 1)x_{k-1} = 0, \tag{6.5.25}$$

which on division by ϕ^2 is

$$\left[\frac{x_{k+1} - 2x_k + x_{k-1}}{\phi^2}\right] + 2\epsilon\left(\frac{x_k - \psi x_{k-1}}{\phi}\right)$$
$$+ \left[\frac{2(1-\psi)x_k + (\phi^2 + \psi^2 - 1)x_{k-1}}{\phi^2}\right]. \tag{6.5.26}$$

Comparison of either Eq. (6.5.21) or (6.5.26) with a standard finite-difference scheme, such as

$$\frac{x_{k+1} - 2x_k + x_{k-1}}{h^2} + 2\epsilon\left(\frac{x_{k+1} - x_{k-1}}{2h}\right) + x_k = 0, \tag{6.5.27}$$

demonstrates that they are clearly "nonstandard."

6.5.3 Duffing Oscillator

The nonlinear Duffing oscillator differential equation is

$$\frac{d^2x}{dt^2} + x + \beta x^3 = 0, \tag{6.5.28}$$

where β is a constant parameter. The first-order system equations are

$$\frac{dx}{dt} = y, \tag{6.5.29}$$

$$\frac{dy}{dt} = -x - \beta x^3. \tag{6.5.30}$$

For this case

$$a = 0, \qquad b = 1, \qquad c = -1, \qquad d = 0, \qquad (6.5.31)$$

$$f(x,y) = 0, \qquad g(x,y) = -\beta x^3, \qquad (6.5.32)$$

and

$$\psi = \cos(h), \qquad \phi = \sin(h). \qquad (6.5.33)$$

The use of the explicit scheme of Eqs. (6.4.1) and (6.4.2) gives

$$\frac{x_{k+1} - \cos(h)x_k}{\sin(h)} = y_k, \qquad (6.5.34)$$

$$\frac{y_{k+1} - \cos(h)y_k}{\sin(h)} = -x_k - \beta x_k^3. \qquad (6.5.35)$$

The elimination of y_k gives

$$\frac{x_{k+1} - 2\cos(h)x_k + \cos^2(h)x_{k-1}}{\sin^2(h)} + x_{k-1} + \beta x_{k-1}^3 = 0, \qquad (6.5.36)$$

which can be rewritten to the form

$$\frac{x_{k+1} - 2x_k + x_{k-1}}{4\sin^2\left(\frac{h}{2}\right)} + x_k + \beta \left[\frac{\sin^2(h)}{4\sin^2\left(\frac{h}{2}\right)}\right] x_{k-1}^3 = 0. \qquad (6.5.37)$$

The corresponding semi-explicit scheme, based on Eqs. (6.4.3) and (6.4.4), is

$$\frac{x_{k+1} - \cos(h)x_k}{\sin(h)} = y_k, \qquad (6.5.38)$$

$$\frac{y_{k+1} - \cos(h)y_k}{\sin(h)} = -x_k - \beta x_{k+1}^3. \qquad (6.5.39)$$

Eliminating y_k and further manipulation of these results gives the expression

$$\frac{x_{k+1} - 2x_k + x_{k-1}}{4\sin^2\left(\frac{h}{2}\right)} + x_k + \beta \left[\frac{\sin^2(h)}{4\sin^2\left(\frac{h}{2}\right)}\right] x_k^3 = 0. \qquad (6.5.40)$$

The question to be asked is which form, Eq. (6.5.37) or Eq. (6.5.40), should be used to calculate numerical solutions to the Duffing differential equation? It has been shown by Mickens [16] that the semi-explicit scheme is the one to use.

(See, also, the arguments presented in Section 5.3.) The basic idea for this choice comes from the fact that the Duffing equation satisfies a conservation law. It follows that all the periodic solutions oscillate with constant amplitude. The semi-explicit scheme of Eq. (6.5.40) has this property, while the explicit scheme, given by Eq. (6.5.37), does not.

6.5.4 $\ddot{x} + x + \epsilon x^2 = 0$

This differential equation arises in the general theory of relativity [17]. Written as a system of first-order equations, it becomes

$$\frac{dx}{dt} = y, \tag{6.5.41}$$

$$\frac{dy}{dt} = -x - \epsilon x^2, \tag{6.5.42}$$

where ϵ is a constant parameter. Based on the result of Section 6.5.3, we will only consider the semi-explicit finite-difference scheme, which for this problem is

$$\frac{x_{k+1} - \cos(h)x_k}{\sin(h)} = y_k, \tag{6.5.43}$$

$$\frac{y_{k+1} - \cos(h)y_k}{\sin(h)} = -x_k - \epsilon x_{k+1}^2. \tag{6.5.44}$$

Eliminating y_k gives

$$\frac{x_{k+1} - 2x_k + x_{k-1}}{\sin^2(h)} + \frac{2[1 - \cos(h)]x_k}{\sin^2(h)} + \epsilon x_k^2 = 0. \tag{6.5.45}$$

Using the fact that

$$2[1 - \cos(h)] = 4\sin^2\left(\frac{h}{2}\right), \tag{6.5.46}$$

we finally obtain

$$\frac{x_{k+1} - 2x_k + x_{k-1}}{4\sin^2\left(\frac{h}{2}\right)} + x_k + \epsilon\left[\frac{\sin^2(h)}{4\sin^2\left(\frac{h}{2}\right)}\right]x_k^2 = 0, \tag{6.5.47}$$

as a nonstandard discrete model for our original differential equation. Again, the arguments of Section 5.3 show that this finite-difference scheme is conservative.

Note that another nonstandard discrete model is given by making the replacement

$$x^2 \rightarrow \frac{x_k(x_{k+1} + x_{k-1})}{2}, \tag{6.5.48}$$

for the nonlinear x^2 term. The finite-difference model in this case is given by the following expression

$$\frac{x_{k+1} - 2x_k + x_{k-1}}{4\sin^2\left(\frac{h}{2}\right)} + x_k + \epsilon\left[\frac{\sin^2(h)}{4\sin^2\left(\frac{h}{2}\right)}\right]\frac{x_k(x_{k+1} + x_{k-1})}{2} = 0. \tag{6.5.49}$$

This is also a conservative scheme that is semi-explicit since x_{k+1} can be calculated in terms of x_k and x_{k-1}. Based on the arguments of Section 5.3, we can conclude that the discrete model of Eq. (6.5.49) is to be preferred over that of Eq. (6.5.47).

6.5.5 van der Pol Oscillator

The van der Pol equation

$$\frac{d^2x}{dt^2} + x = \epsilon(1 - x^2)\frac{dx}{dt}, \tag{6.5.50}$$

can be written in system form as

$$\frac{dx}{dt} = y, \tag{6.5.51}$$

$$\frac{dy}{dt} = -x + \epsilon(1 - x^2)y, \tag{6.5.52}$$

where ϵ is a positive parameter. Note that the linear terms of these equations are

$$\frac{du}{dt} = w, \tag{6.5.53}$$

$$\frac{dw}{dt} = -u + \epsilon y, \tag{6.5.54}$$

and correspond to an unstable "damped" linear oscillator. (See Eqs. (6.5.14) and (6.5.15).) For this case

$$a = 0, \quad b = 1, \quad c = -1, \quad d = \epsilon, \tag{6.5.55}$$

and the functions ϕ and ψ are given by the expressions

$$\psi(\epsilon, h) = -\frac{\epsilon e^{\epsilon h/2}}{2\sqrt{1-\frac{\epsilon^2}{4}}} \cdot \sin\left(\sqrt{1-\frac{\epsilon^2}{4}}\right) h + e^{\epsilon k/2} \cos\left(\sqrt{1-\frac{\epsilon^2}{4}}\right) h, \qquad (6.5.56)$$

$$\phi(\epsilon, h) = \frac{e^{\epsilon h/2}}{\sqrt{1-\frac{\epsilon^2}{4}}} \cdot \sin\left(\sqrt{1-\frac{\epsilon^2}{4}}\right) h. \qquad (6.5.57)$$

Therefore, the semi-explicit scheme for the van der Pol differential equation is

$$\frac{x_{k+1} - \psi x_k}{\phi} = y_k, \qquad (6.5.58)$$

$$\frac{y_{k+1} - \psi y_k}{\phi} = -x_k + \epsilon y_k - \epsilon x_{k+1}^2 y_k. \qquad (6.5.59)$$

Eliminating y_k and rewriting the resulting expression gives

$$\frac{x_{k+1} - 2x_k + x_{k-1}}{\phi^2} + \frac{2(1-\psi)x_k + (\psi^2 + \phi^2 - 1)x_{k-1}}{\phi^2}$$
$$= \epsilon(1 - x_k^2)\left[\frac{x_k - \psi x_{k-1}}{\phi}\right]. \qquad (6.5.60)$$

Another possibility for constructing a discrete model of the van der Pol equation is to consider the following set of linear terms

$$\frac{du}{dt} = w, \qquad (6.5.61)$$

$$\frac{dw}{dt} = -u. \qquad (6.5.62)$$

For this case, the linear term ϵy is incorporated into the function $g(x, y) = \epsilon(1-x^2)y$. If we do this, the functions ψ and ϕ become

$$\psi = \cos(h), \qquad \phi = \sin(h) \qquad (6.5.63)$$

and, the semi-explicit scheme is

$$\frac{x_{k+1} - \cos(h)x_k}{\sin(h)} = y_k, \qquad (6.5.64)$$

$$\frac{y_{k+1} - \cos(h)y_k}{\sin(h)} = -x_k + \epsilon(1 - x_{k+1}^2)y_k. \tag{6.5.65}$$

Eliminating y_k gives

$$\frac{x_{k+1} - 2x_k + x_{k-1}}{4\sin^2\left(\frac{h}{2}\right)} + x_k = \epsilon\left[\frac{\sin(h)}{2\sin\left(\frac{h}{2}\right)}\right](1 - x_k^2)\left[\frac{x_k - \cos(h)x_{k-1}}{2\sin\left(\frac{h}{2}\right)}\right]. \tag{6.5.66}$$

Two things should be noted about this last relation. First, it is similar to one of the discrete models investigated in Section 5.4. Second, this scheme does not have the correct linear stability properties: The van der Pol differential equation and the finite-difference equation, given by Eq. (6.5.60), both have an unstable fixed-point $(\bar{x}, \bar{y}) = (0,0)$; the scheme of Eq. (6.5.66) has neutral stability, i.e., the local stability properties of the harmonic oscillator. Thus, we conclude that Eq. (6.5.60) should be used as a discrete model for the van der Pol differential equation.

6.5.6 Lewis Oscillator

The differential equation for the nonlinear Lewis oscillator is [7]

$$\frac{d^2x}{dt^2} + x = \epsilon(1 - |x|)\frac{dy}{dt}, \tag{6.5.67}$$

where ϵ is a positive parameter. The corresponding system equations are

$$\frac{dx}{dt} = y, \tag{6.5.68}$$

$$\frac{dy}{dt} = -x + \epsilon y - \epsilon|x|y. \tag{6.5.69}$$

Since the linear terms of these equations are exactly the same as for the van der Pol differential equation, the functions ϕ and ψ for the Lewis oscillator are also given by Eqs. (6.5.56) and (6.5.57). Thus, the semi-explicit scheme is

$$\frac{x_{k+1} - \psi x_k}{\phi} = y_k, \tag{6.5.70}$$

$$\frac{y_{k+1} - \psi y_k}{\phi} = -x_k + \epsilon y_k - \epsilon|x_{k+1}|y_k, \tag{6.5.71}$$

or upon rewriting

$$\frac{x_{k+1} - 2x_k + x_{k-1}}{\phi^2} + \frac{2(1 - \psi)x_k + (\psi^2 + \phi^2 - 1)x_{k-1}}{\phi^2}$$
$$= \epsilon(1 - |x_k|)\left[\frac{x_k - \psi x_{k-1}}{\phi}\right]. \tag{6.5.72}$$

6.5.7 General Class of Nonlinear Oscillators

Section 5.5 presented and discussed the construction of a nonstandard finite-difference scheme for a general class of nonlinear oscillators for which the equation of motion is

$$\frac{d^2x}{dt^2} + \dot{x} + f(x^2)\frac{dx}{dt} + g(x^2)x = 0. \tag{6.5.73}$$

Written in system form, this equation becomes

$$\frac{dx}{dt} = y, \tag{6.5.74}$$

$$\frac{dy}{dt} = -x - g(x^2)x - f(x^2)y. \tag{6.5.75}$$

It is assumed that the functions $f(x^2)$ and $g(x^2)$ have the following properties:

$$f(x^2) = f_0 + f_1 x^2 + \bar{f}(x^2), \tag{6.5.76}$$

$$\bar{f}(x^2) = O(x^4), \tag{6.5.77}$$

$$g(0) = 0. \tag{6.5.78}$$

Consequently, the linear parts of Eqs. (6.5.74) and (6.5.75) are

$$\frac{du}{dt} = w, \tag{6.5.79}$$

$$\frac{dw}{dt} = -u - f_0 w, \tag{6.5.80}$$

and ψ and ϕ, in Eqs. (6.3.11) and (6.3.12), are to be calculated using

$$a = 0, \quad b = 1, \quad c = -1, \quad d = -f_0. \tag{6.5.81}$$

Therefore, the semi-explicit scheme for Eqs. (6.5.74) and (6.5.75) is

$$\frac{x_{k+1} - \psi x_k}{\phi} = y_k, \tag{6.5.82}$$

$$\frac{y_{k+1} - \psi y_k}{\phi} = -x_k - f_0 y_k - [f_1 x_{k+1}^2 + \bar{f}(x_{k+1}^2)]y_k - g(x_{k+1}^2)x_{k+1}. \tag{6.5.83}$$

Finally, eliminating y_k and rearranging the various terms gives

$$\left[\frac{x_{k+1} - 2x_k + x_{k-1}}{\phi^2}\right] + \left[\frac{2(1 - \psi)x_k + (\psi^2 + \phi^2 - 1)x_{k-1}}{\phi^2}\right]$$

$$+ f(x_k^2)\left[\frac{x_k - \psi x_{k-1}}{\phi}\right] + g(x_k^2)x_k = 0. \tag{6.5.84}$$

Combining this scheme with the nonlocal symmetric modeling of the $g(x^2)x$ term gives the following discrete representation

$$\left[\frac{x_{k+1} - 2x_k + x_{k-1}}{\phi^2}\right] + \frac{2(1 - \psi)x_k + (\psi^2 + \phi^2 - 1)x_{k-1}}{\phi^2}$$

$$+ f(x_k^2)\left[\frac{x_k - \psi x_{k-1}}{\phi}\right]$$

$$+ g(x_k^2)\left(\frac{x_{k+1} + x_{k-1}}{2}\right) = 0. \tag{6.5.85}$$

Note that in contrast to Eq. (5.5.4), the discrete models, given in Eqs. (6.5.84) and (6.5.85), have the correct linear stability properties.

6.5.8 Batch Fermentation Processes

The differential equations are given by Eqs. (6.1.7) and (6.1.8). The linear terms correspond to the equations

$$\frac{du}{dt} = -(A\alpha\beta)w, \tag{6.5.86}$$

$$\frac{dw}{dt} = (B\alpha\beta)u + (B\gamma\beta^2)w, \tag{6.5.87}$$

where

$$a = 0, \qquad b = -A\alpha\beta, \qquad c = B\alpha\beta, \qquad d = B\gamma\beta^2. \tag{6.5.88}$$

With these values, the functions ϕ and ψ can be determined using Eqs. (6.3.11) and (6.3.12). This gives the following semi-explicit finite-difference scheme

$$\frac{x_{k+1} - \psi x_k}{\phi} = -(A\alpha\beta)y_k + [(A\alpha)x_k y_k + (A\beta)y_k^2] - Ax_k y_k^2, \qquad (6.5.89)$$

$$\begin{aligned}\frac{y_{k+1} - \psi y_k}{\phi} = {}& [(B\alpha\beta)x_k + (B\gamma\beta^2)y_k] \\ & + [(\alpha B)x_{k+1}^2 + (2B\beta\gamma)x_{k+1}y_k - (B\beta^2\epsilon)y_k^2] \\ & + [(B\gamma)x_{k+1}^2 y_k - (2B\beta\epsilon)x_{k+1}y_k^2] - (B\epsilon)x_{k+1}^2 y_k^2. \quad (6.5.90)\end{aligned}$$

6.6 Summary

In this chapter, we have shown that it is possible to construct finite-difference schemes for two coupled, first-order nonlinear differential equations, for which there is only a single (real) fixed-point, such that the difference equations have exactly the same linear stability properties as the differential equations for all finite values of the step-size. This result is very important since standard finite-difference schemes do not have this property. In addition, this result implies that elementary numerical instabilities do not occur.

Based on the earlier work of Mickens [4, 16, 18], it was concluded that the semi-explicit procedure, given by Eqs. (6.4.3) and (6.4.4), is the proper discrete modeling technique to use. The semi-explicit scheme is an explicit method for which the variables are calculated in a definite order: first x_{k+1} is determined and then y_{k+1}.

As in the previous work of Mickens [2, 3, 4, 18], it was found that generalized representations of discrete derivatives appeared in the nonstandard finite-difference schemes constructed in this chapter; see, for example, Eqs. (6.3.13) and (6.3.14). This feature is ubiquitous in the construction of nonstandard discrete models of differential equations.

The semi-explicit scheme is not an exact finite-difference discrete model. It is a *best finite-difference scheme*. This means that it was constructed in such a way that a critical feature of its solution corresponded exactly to the related property of the original differential equation. In this case, the critical property was the nature of the stability for the fixed-point at $(\bar{x}, \bar{y}) = (0,0)$.

References

1. F. B. Hildebrand, *Finite-Difference Equations and Simulations* (Prentice-Hall; Englewood Cliffs, NJ; 1968). Sections 2.6, 2.8 and 2.10.

2. R. E. Mickens, *Numerical Methods for Partial Differential Equations* 5, 313–325 (1989). Exact solutions to a finite-difference model for a nonlinear reaction-advection equation: Implications for numerical analysis.

3. R. E. Mickens and A. Smith, *Journal of the Franklin Institute,* 327, 143–149 (1990). Finite-difference models of ordinary differential equations: Influence of denominator functions.

4. R. E. Mickens, *Dynamic System and Applications* 1, 329–340 (1992). Finite-difference schemes having the correct linear stability properties for all finite step-sizes II.

5. B. van der Pol, *Philosophical Magazine* 43, 177–193 (1922). On a type of oscillation hysteresis in a simple triode generator.

6. R. E. Mickens, *Nonlinear Oscillations* (Cambridge University Press, New York, 1981).

7. J. B. Lewis, *Transactions of the American Institute of Electrical Engineering, Part II* 72, 449–453 (1953). The use of nonlinear feedback to improve the transient response of a servomechanism.

8. R. E. Mickens and I. Ramadhani, *Journal of Sound and Vibration* 154, 190–193 (1992). Investigation of an anti-symmetric quadratic nonlinear oscillator.

9. Y. Lenbury, P. S. Crooke and R. D. Tanner, *BioSystems* 19, 15–22 (1986). Relating damped oscillations to the sustained limit cycles describing real and ideal batch fermentation processes.

10. L. Edelstein-Keshet, *Mathematical Models in Biology* (Random House/ Birkhäuser, New York, 1988).

11. O. Sporns and F. F. Seelig, *BioSystems* **19**, 83–89 (1986). Oscillations in theoretical models of induction.

12. R. J. Field and M. Burger, editors, *Oscillating and Traveling Waves in Chemical Systems* (Wiley-Interscience, New York, 1985).

13. D. Potter, *Computational Physics* (Wiley-Interscience, New York, 1973).

14. J. M. T. Thompson and H. B. Stewart, *Nonlinear Dynamics and Chaos* (Wiley, New York, 1986).

15. S. L. Ross, *Differential Equations* (Xerox; Lexington, MA; 1974, 2nd edition). Chapter 7.

16. R. E. Mickens, *Journal of Sound and Vibration* **124**, 194–198 (1988). Properties of finite-difference models of nonlinear conservative oscillators.

17. W. Rindler, *Essential Relativity: Special, General and Cosmological* (Van Nostrand Reinhold, New York, 1969), section 7.5.

18. R. E. Mickens, Investigation of finite-difference models of the van der Pol equation, in *Differential Equations and Applications*, A. R. Aftabizadeh, editor (Ohio University Press; Columbus, OH; 1988), pp. 210–215.

Chapter 7
PARTIAL DIFFERENTIAL EQUATIONS

7.1 Introduction

Partial differential equations provide valuable mathematical models for dynamical systems that involve both space and time variables [1–8]. In this chapter, we study the construction of nonstandard finite-difference schemes for a number of linear and nonlinear partial differential equations. In general, these equations are first-order in the time derivative and first- or second-order in the space derivatives. These equations include various one space dimension modifications of wave, diffusion and Burgers' partial differential equations. The nonlinearities considered are generally quadratic functions of the dependent variable and its derivatives. The use of only quadratic nonlinear terms follows from the result that for these expressions exact special solutions can often be found for the partial differential equation under study. These special solutions can then be used in the construction of nonstandard discrete models. However, it should be noted that exact finite-difference schemes are not expected to exist for partial differential equations [9, 10]. This is a general consequence of the realization that for any arbitrarily specified partial differential equation, no precise definition of the general solution can be given [11].

For each partial differential equation considered, a comparison will be made to the standard finite-difference schemes and how the solutions of the various nonstandard and nonstandard discrete models differ from each other. The results obtained in this chapter rely heavily on the concept of "best" finite-difference scheme as introduced in Chapter 3. In summary, a *best scheme* is a discrete model of a differential equation that incorporates as many of the properties of the differential equation as possible into its mathematical structure. While best schemes correspond to nonstandard discrete models, they are not, in general, unique. Additional information or requirements are usually needed to obtain uniqueness.

Sections 7.2, 7.3 and 7.4 treat, respectively, the discrete modeling of wave, diffusion and Burgers' type partial differential equations. In Section 7.5, we summarize what has been found for the various finite-difference schemes and carry out a general discussion on the problems of constructing better discrete models for partial differential equations.

7.2 Wave Equations

7.2.1 $u_t + u_x = 0$

The unidirectional wave equation

$$u_t + u_x = 0, \tag{7.2.1}$$

treated as an initial value problem, i.e.,

$$u(x,0) = f(x) = \text{given}, \tag{7.2.2}$$

has the exact solution

$$u(x,t) = f(x-t). \tag{7.2.3}$$

This corresponds to a wave form traveling with unit velocity to the right.

A direct calculation [12] shows that the partial difference equation

$$u_m^{k+1} = u_{m-1}^k, \tag{7.2.4}$$

has the exact solution

$$u_m^k = h(m-k) \tag{7.2.5}$$

where $h(z)$ is an arbitrary function of z. For

$$\Delta x = \Delta t = h, \tag{7.2.6}$$

$$t_k = hk, \qquad x_m = hm, \tag{7.2.7}$$

we have

$$u_m^k = h\left(\frac{x_m - t_k}{h}\right) = H(x_m - t_k). \tag{7.2.8}$$

Thus, it can be concluded that Eq. (7.2.4) is an exact finite-difference model of the unidirectional wave equation given by Eq. (7.2.1), i.e.,

$$u_m^k = u(x_m, t_k), \tag{7.2.9}$$

where u_m^k is a solution of Eq. (7.2.4) and $u(x,t)$ is the corresponding solution of Eq. (7.2.1).

The above partial difference equation can be rewritten as

$$\frac{u_m^{k+1} - u_m^k}{\psi(\Delta t)} + \frac{u_m^k - u_{m-1}^k}{\psi(\Delta x)} = 0, \qquad \Delta t = \Delta x, \tag{7.2.10}$$

where $\psi(z)$ has the property

$$\psi(z) = z + O(z^2). \tag{7.2.11}$$

The denominator function $\psi(z)$ is not determined by this analysis. Any $\psi(z)$ that satisfies the condition given in Eq. (7.2.11) will work. The simplest choice is $\psi(z) = z$. However, as Eq. (7.2.4) indicates, for the unidirectional wave equation, the particular choice is irrelevant since $\Delta t = \Delta x$ and the denominator functions drop out of the calculations. Note that the discrete time-derivative is forward Euler, while the discrete space-derivative is a backward Euler.

7.2.2 $u_t - u_x = 0$

There is a second linear unidirectional wave equation that describes a wave form traveling to the left with unit velocity. It is given by the equation

$$u_t - u_x = 0. \tag{7.2.12}$$

For the initial value problem

$$u(x,0) = g(x) = \text{given}, \tag{7.2.13}$$

the exact solution is

$$u(x,t) = g(x+t). \tag{7.2.14}$$

The partial difference equation

$$u_m^{k+1} = u_{m+1}^k, \tag{7.2.15}$$

has the exact solution [12]

$$u_m^k = p(m+k). \tag{7.2.16}$$

Using the conditions of Eqs. (7.2.6) and (7.2.7), we have

$$u_m^k = P(x_m + t_k), \tag{7.2.17}$$

and, consequently, conclude that Eq. (7.2.15) is an exact finite-difference model for Eq. (7.2.12).

Proceeding as in the last section, Eq. (7.2.15) can be rewritten to the form

$$\frac{u_m^{k+1} - u_m^k}{\psi(\Delta t)} - \frac{u_{m+1}^k - u_m^k}{\psi(\Delta x)} = 0, \tag{7.2.18}$$

where $\psi(z)$ has the property given by Eq. (7.2.11) and the condition

$$\Delta t = \Delta x, \tag{7.2.19}$$

must be satisfied. For this discrete model, both discrete first-derivatives are given by forward-Euler representations.

7.2.3 $u_{tt} - u_{xx} = 0$

The full or dual direction wave equation is

$$u_{tt} - u_{xx} = 0. \tag{7.2.20}$$

Its general solution is

$$u(x,t) = f(x-t) + g(x+t), \tag{7.2.21}$$

where $f(z)$ and $g(z)$ are arbitrary functions having second derivatives [4]. The exact finite-difference equation for the wave equation is [4]

$$u_m^{k+1} + u_m^{k-1} = u_{m+1}^k + u_{m-1}^k. \qquad (7.2.22)$$

To prove this, let us show that

$$w_m^k = F(m-k) + G(m+k) \qquad (7.2.23)$$

is a solution to Eq. (7.2.22). Note that

$$u_m^{k+1} = F(m-k-1) + G(m+k+1), \qquad (7.2.24)$$

$$u_m^{k-1} = F(m-k+1) + G(m+k-1), \qquad (7.2.25)$$

and

$$u_{m+1}^k = F(m-k+1) + G(m+k+1), \qquad (7.2.26)$$

$$u_{m-1}^k = F(m-k+1) + G(m+k+1). \qquad (7.2.27)$$

Substitution of these expressions into, respectively, the left- and right-sides of Eq. (7.2.22) gives the desired result, namely Eq. (7.2.23) is the general solution.

Subtracting $2u_m^k$ from both sides of Eq. (7.2.22) and dividing by $\phi(h)$ where

$$\phi(x) = h^2 + O(h^4), \qquad (7.2.28)$$

gives

$$\frac{u_m^{k+1} - 2u_m^k + u_m^{k-1}}{\phi(\Delta t)} = \frac{u_{m+1}^k - 2u_m^k + u_{m-1}^k}{\phi(\Delta x)}, \qquad (7.2.29)$$

where the condition

$$\Delta t = \Delta x \qquad (7.2.30)$$

is required. Note that the exact analytical expression for $\phi(h)$ is not needed since, with the condition of Eq. (7.2.30), the denominator functions drop out of the calculation.

7.2.4 $u_t + u_x = u(1 - u)$

The exact finite-difference scheme for this nonlinear reaction-advection equation has already been given and discussed in Section 3.3 [9]. It is given by the expression

$$\frac{u_m^{k+1} - u_m^k}{\psi(\Delta t)} + \frac{u_m^k - u_{m-1}^k}{\psi(\Delta x)} = u_{m-1}^k(1 - u_m^{k+1}), \qquad (7.2.31)$$

where the denominator function is

$$\psi(h) = e^h - 1, \qquad (7.2.32)$$

and the following requirement must be satisfied

$$\Delta t = \Delta x. \qquad (7.2.33)$$

The following points should be observed:

(i) The discrete derivatives correspond exactly to those found previously for the unidirectional wave equation, the linear terms of this nonlinear differential equation.

(ii) The denominator function $\psi(h)$ has a specified form given by Eq. (7.2.32). This is a consequence of the nonlinearity of the partial differential equation.

(iii) There is a functional relation between the step-sizes, i.e., $\Delta t = \Delta h$.

(iv) The nonlinear u^2 term is modeled nonlocally on the discrete space-time grid, i.e.,

$$u^2 \rightarrow u_{m-1}^k u_m^{k+1}. \qquad (7.2.34)$$

(iv) Standard finite-difference schemes, such as

$$\frac{u_m^{k+1} - u_m^k}{\Delta t} + \frac{u_{m+1}^k - u_m^k}{\Delta x} = u_m^k(1 - u_m^k), \qquad (7.2.35)$$

do not have these features and, consequently, are expected to have numerical instabilities. (See references [8, 9, 12] and Section 2.5.)

7.2.5 $u_t + u_x = bu_{xx}$

The linear advection-diffusion equation

$$u_t + u_x = bu_{xx}, \qquad b > 0, \qquad\qquad (7.2.36)$$

plays a very important role in the analysis of certain physical phenomena in fluid dynamics [2, 3]. Also, of equal importance is that this partial differential equation provides a good model for testing finite-difference schemes constructed for the numerical integration of more complicated equations. A vast literature exist on the detailed analysis of the stability properties for these finite-difference schemes. These include the works of Peyret and Taylor [13], Chan [14], Strikewerda [15], and Bentley, Pinder and Herrera [16].

Two standard finite-difference schemes that have been investigated in detail with regard to their stability properties are

$$\frac{u_m^{k+1} - u_m^k}{\Delta t} + \frac{u_{m+1}^k - u_{m-1}^k}{\Delta x} = b\left[\frac{u_{m+1}^k - 2u_m^k + u_{m-1}^k}{(\Delta x)^2}\right], \qquad (7.2.37)$$

and

$$\frac{u_m^{k+1} - u_m^k}{\Delta t} + \frac{u_m^k - u_{m-1}^k}{\Delta x} = b\left[\frac{u_{m+1}^k - 2u_m^k + u_{m-1}^k}{(\Delta x)^2}\right]. \qquad (7.2.38)$$

It should be clear that both of these schemes will give rise to numerical instabilities. The first because it models the discrete space-derivative by a second-order central difference scheme and also because there is no relationship between the two step-sizes, and the second because of the absence of a functional relation between Δt and Δx. (However, the requirements of stability do give a relationship between these two step-sizes [13–16].)

Our goal is to construct a conditionally stable explicit nonstandard finite-difference model for the linear advection-diffusion equation [10]. (In brief, a conditional stable model is one for which the discrete-time dependency of the solution is bounded as $k \to \infty$. For details, see references [13, 14, 15, 17].) To begin, we note that Eq. (7.2.36) can be decomposed into the following two sub-equations

$$u_t + u_x = 0, \qquad (7.2.39)$$

$$u_x = bu_{xx}. \qquad (7.2.40)$$

Each of these differential equations has an exact discrete model. They are given, respectively, by the expression [18, 19]

$$\frac{u_m^{k+1} - u_m^k}{\psi(\Delta t)} + \frac{u_m^k - u_{m-1}^k}{\psi(\Delta x)} = 0, \qquad (7.2.41a)$$

$$\psi(h) = h + O(h^2), \qquad \Delta x = \Delta t, \qquad (7.2.41b)$$

$$\frac{u_m - u_{m-1}}{\Delta x} = b \left[\frac{u_{m+1} - 2u_m + u_{m-1}}{b(e^{\Delta x/b} - 1)\Delta x} \right]. \qquad (7.2.42)$$

A finite-difference model for Eq. (7.2.36) that combines both of these discrete sub-equations is

$$\frac{u_m^{k+1} - u_m^k}{\Delta t} + \frac{u_m^k - u_{m-1}^k}{\Delta x} = b \left[\frac{u_{m+1}^k - 2u_m^k + u_{m-1}^k}{b(e^{\Delta x/b} - 1)\Delta x} \right]. \qquad (7.2.43)$$

For this equation, we do not know what the relation is between the step-sizes Δx and Δt. To find this relation, we proceed as follows. First, Eq. (7.2.43) can be solved for u_m^{k+1} to give

$$u_m^{k+1} = \beta u_{m+1}^k + (1 - \alpha - 2\beta)u_m^k + (\alpha + \beta)u_{m-1}^k, \qquad (7.2.44)$$

where

$$\alpha = \frac{\Delta t}{\Delta x}, \qquad \beta = \frac{\alpha}{(e^{\Delta x/b} - 1)}. \qquad (7.2.45)$$

The conditional stability of the solutions to either Eq. (7.2.43) or (7.2.44) can be assured by the use of a result due to Forsythe and Wasow [17]. (See also reference [20].) They proved that if all the coefficients on the right-side of Eq. (7.2.44) are non-negative, then the finite-difference scheme is (conditionally) stable. This condition is related to the requirement that the solutions to Eq. (7.2.36) satisfy a min-max principle. However, in general, the solutions to Eq. (7.2.44) do not satisfy this principle. The imposing of this min-max principle on the solutions to Eq. (7.2.44) gives the conditional stability requirement.

The coefficients of the u_{m+1}^k and u_{m-1}^k terms are positive by definition of α and β, and the fact that $b \geq 0$. Hence, the conditional stability requirement is

$$1 - \alpha - 2\beta \geq 0, \tag{7.2.46}$$

or

$$\Delta t \leq \Delta x \left[\frac{e^{\Delta x/b} - 1}{e^{\Delta x/b} + 1} \right]. \tag{7.2.47}$$

This inequality places a restriction on the time step-size once the space step-size is selected. By choosing the equality sign, we have a functional relation between Δx and Δt. Note that this relation

$$\Delta t = \Sigma(\Delta x, b) = \Delta x \left[\frac{e^{\Delta x/b} - 1}{e^{\Delta x/b} + 1} \right], \tag{7.2.48}$$

also depends on the parameter b.

In summary, the linear advection-diffusion equation has a (nonstandard) best finite-difference scheme given by Eq. (7.2.43) where the step-sizes are related by the result expressed in Eq. (7.2.48).

7.3 Diffusion Equations

The construction of better finite-difference models for the linear diffusion equation

$$u_t = bu_{xx}, \qquad b > 0, \tag{7.3.1}$$

has been studied since the beginning of modern numerical analysis [4, 6, 7, 15, 17, 20, 21, 22, 23]. The simplest scheme is the standard explicit forward-Euler which is given by the expression

$$\frac{u_m^{k+1} - u_m^k}{\Delta t} = b \left[\frac{u_{m+1}^k - 2u_m^k + u_{m-1}^k}{(\Delta x)^2} \right]. \tag{7.3.2}$$

The conditional stability requirement is [6]

$$\Delta \le \frac{(\Delta x)^2}{2b}. \tag{7.3.3}$$

This section will be devoted to an investigation of how nonstandard finite-difference schemes can be constructed for diffusion type partial differential equations.

7.3.1 $u_t = au_{xx} + bu$

Consider the linear diffusion equation

$$u_t = au_{xx} + bu, \tag{7.3.4}$$

where a and b are constants with $a \ge 0$. The two sub-equations

$$\frac{du}{dt} = bu, \tag{7.3.5}$$

$$a\frac{d^2u}{dx^2} + bu = 0, \tag{7.3.6}$$

both have exact finite-difference schemes. They are

$$\frac{u^{k+1} - u^k}{\left(\frac{e^{b\Delta t}-1}{b}\right)} = bu^k, \tag{7.3.7}$$

$$a\left\{\frac{u_{m+1} - 2u_m + u_{m-1}}{\left(\frac{4a}{b}\right)\sin^2\left[\sqrt{\frac{b}{a}}\left(\frac{\Delta x}{2}\right)\right]}\right\} + bu_m = 0. \tag{7.3.8}$$

(Note that these results are correct for any value of the sign for b.) There are two ways of combining Eqs. (7.3.7) and (7.3.8) to form a discrete model for the full linear diffusion equation. The first is the explicit scheme

$$\frac{u_m^{k+1} - u_m^k}{\left(\frac{e^{b\Delta t}-1}{b}\right)} = a\left\{\frac{u_{m+1}^k - 2u_m^k + u_{m-1}^k}{\left(\frac{4a}{b}\right)\sin^2\left[\sqrt{\frac{b}{a}}\left(\frac{\Delta x}{2}\right)\right]}\right\} + bu_m^k, \tag{7.3.9}$$

the second is the implicit scheme

$$\frac{u_m^{k+1} - u_m^k}{\left(\frac{e^{b\Delta t} - 1}{b}\right)} = a\left\{\frac{u_{m+1}^{k+1} - 2u_m^{k+1} + u_{m-1}^{k+1}}{\left(\frac{4a}{b}\right)\sin^2\left[\sqrt{\frac{b}{a}}\left(\frac{\Delta x}{2}\right)\right]}\right\} + bu_m^k. \tag{7.3.10}$$

Observe that in both schemes, the bu term is evaluated at the k-th discrete-time step rather than the $(k + 1)$-th step. Also, both schemes reduce to the correct discrete model of the corresponding sub-equations.

Finite-difference schemes for the simple diffusion equation are obtained by taking the limit $b \to 0$. Doing this for Eq. (7.3.9) gives the standard explicit model of Eq. (7.3.2), while Eq. (7.3.10) goes to the standard implicit form

$$\frac{u_m^{k+1} - u_m^k}{\Delta t} = a\left[\frac{u_{m+1}^{k+1} - 2u_m^{k+1} + u_{m-1}^{k+1}}{(\Delta x)^2}\right]. \tag{7.3.11}$$

7.3.2 $u_t = uu_{xx}$

The nonlinear diffusion equation

$$u_t = uu_{xx} \tag{7.3.12}$$

has the special rational solution

$$u(x, t) = \frac{-\left(\frac{\alpha}{2}\right)x^2 + \beta_2 x + \beta_2}{\alpha_1 + \alpha t}. \tag{7.3.13}$$

This can be shown by using the method of separation of variables and writing $u(x, t)$ as

$$u(x, t) = X(x)T(t), \tag{7.3.14}$$

and substituting this into Eq. (7.3.12) to obtain

$$\frac{1}{T^2}\frac{dT}{dt} = \frac{d^2X}{dx^2} = -\alpha, \tag{7.3.15}$$

where α is the separation constant. The differential equations

$$\frac{dT}{dt} = -\alpha T^2, \qquad \frac{d^2X}{dx^2} = -\alpha, \tag{7.3.16}$$

have the respective solutions

$$T(t) = \frac{1}{\alpha_1 + \alpha t},$$

(7.3.17)

$$X(x) = -\left(\frac{\alpha}{2}\right)x^2 + \beta_1 x + \beta_2,$$

(7.3.18)

where $(\alpha_1, \beta_1, \beta_2)$ are arbitrary integration constants.

Based on the nonstandard modeling rules, given in Section 3.4, the nonlinear term uu_{xx}, on the right-side of Eq. (7.3.12), must be modeled by a discrete form that is nonlocal in the time variable. The simplest two choices are

$$uu_{xx} \rightarrow \begin{cases} u_m^{k+1}\left[\frac{u_{m+1}^k - 2u_m^k + u_{m-1}^k}{\phi(\Delta x)}\right], \\ u_m^k\left[\frac{u_{m+1}^{k+1} - 2u_m^{k+1} + u_{m-1}^{k+1}}{\phi(\Delta x)}\right], \end{cases}$$

(7.3.19)

where the denominator function ϕ has the property

$$\phi(h) = h^2 + O(h^4).$$

(7.3.20)

These lead to the following two finite-difference schemes

$$\frac{u_m^{k+1} - u_m^k}{\psi(\Delta t)} = u_m^{k+1}\left[\frac{u_{m+1}^k - 2u_m^k + u_{m-1}^k}{\phi(\Delta x)}\right],$$

(7.3.21)

$$\frac{u_m^{k+1} - u_m^k}{\psi(\Delta t)} = u_m^k\left[\frac{u_{m+1}^{k+1} - 2u_m^{k+1} + u_{m-1}^{k+1}}{\phi(\Delta x)}\right],$$

(7.3.22)

where

$$\psi(h) = h + O(h^2).$$

(7.3.23)

These models are, respectively, explicit and implicit finite-difference schemes for Eq. (7.3.12).

Consider first Eq. (7.3.21). A special exact solution can be found by the method of separation of variables [12], i.e., take u_m^k to be

$$u_m^k = C^k D_m,$$

(7.3.24)

where C^k is a function only of the discrete-time k and D_m depends only on the discrete-space variable m. Substitution of Eq. (7.3.24) into Eq. (7.3.21) gives

$$\frac{(C^{k+1} - C^k)D_m}{\psi} = C^{k+1}C^k D_m \left[\frac{D_{m+1} - 2D_m + D_{m-1}}{\phi}\right] \qquad (7.3.25)$$

and

$$\frac{C^{k+1} - C^k}{\psi C^{k+1}C^k} = \frac{D_{m+1} - 2D_m + D_{m-1}}{\phi} = -\alpha, \qquad (7.3.26)$$

where α is the separation constant. The two ordinary difference equations

$$C^{k+1} - C^k = -\alpha\psi C^{k+1}C^k, \qquad (7.3.27)$$

$$D_{m+1} - 2D_m + D_{m-1} = -\alpha\phi, \qquad (7.3.28)$$

have the respective solutions [12]

$$C^k = \frac{1}{A_1 + \alpha\psi k}, \qquad (7.3.29)$$

$$D_m = -\left(\frac{\alpha}{2}\right)\phi m^2 + B_1 m + B_2, \qquad (7.3.30)$$

where (A_1, B_1, B_2) are arbitrary constants. Comparison of Eqs. (7.3.17) and (7.3.29), and (7.3.18) and (7.3.30) shows that if we require for the special rational solution

$$u_m^k = u(x_m, t_k), \qquad (7.3.31)$$

then we must have

$$\psi(\Delta t) = \Delta t, \qquad \phi(\Delta x) = (\Delta x)^2, \qquad (7.3.32)$$

and

$$A_1 = \alpha_1, \qquad B_1 = (\Delta x)\beta_1, \qquad B_2 = \beta_2. \qquad (7.3.33)$$

Thus, we conclude that the explicit finite-difference scheme

$$\frac{u_m^{k+1} - u_m^k}{\Delta t} = u_m^{k+1}\left[\frac{u_{m+1}^k - 2u_m^k + u_{m-1}^k}{(\Delta x)^2}\right], \qquad (7.3.34)$$

178

has exactly the same rational solution as the original nonlinear diffusion equation given by Eq. (7.3.13). The same set of steps shows that the implicit scheme of Eq. (7.3.22) also has this property provided that the denominator functions $\psi(\Delta t)$ and $\phi(\Delta x)$ are those of Eq. (7.3.32). This implicit scheme is

$$\frac{u_m^{k+1} - u_m^k}{\Delta t} = u_m^k \left[\frac{u_{m+1}^{k+1} - 2u_m^{k+1} + u_{m-1}^{k+1}}{(\Delta x)^2}\right]. \tag{7.3.35}$$

With no other restrictions, we cannot choose between these discrete models. However, experience with the standard techniques of numerical analysis indicates that the implicit scheme should provide "better" numerical results [6, 7].

It should be clear that a standard finite-difference model, such as

$$\frac{u_m^{k+1} - u_m^k}{\Delta t} = u_m^k \left[\frac{u_{m+1}^k - 2u_m^k + u_{m-1}^k}{(\Delta x)^2}\right], \tag{7.3.36}$$

cannot have the exact rational solution of Eqs. (7.3.29) and (7.3.30). The separation of variables form $u_m^k = C^k D_m$ gives for C^k and D_m the equations

$$\frac{(C^{k+1} - C^k)D_m}{\Delta t} = (C^k)^2 D_m \left[\frac{D_{m+1} - 2D_m + D_{m-1}}{(\Delta x)^2}\right] \tag{7.3.37}$$

and

$$C^{k+1} - C^k = -\alpha(\Delta t)(C^k)^2, \tag{7.3.38}$$

$$D_{m+1} - 2D_m + D_{m-1} = -\alpha(\Delta x)^2, \tag{7.3.39}$$

where α is the separation constant. The equation for C^k is the Logistic difference equation and has a variety of solution behaviors, none of which are given by Eq. (7.3.29). Hence, the discrete model of Eq. (7.3.36) will have numerical instabilities.

Finally, it should be observed that at this stage of the investigation no relationship exists between the two step-sizes, Δx and Δt.

7.3.3 $u_t = uu_{xx} + \lambda u(1 - u)$

The nonlinear diffusion equation [24]

$$u_t = uu_{xx} + \lambda u(1 - u) \tag{7.3.40}$$

can be decomposed into three special limiting cases. They are (i) the space-independent equation

$$u_t = \lambda u(1 - u),\qquad(7.3.41)$$

(ii) the time-independent equation

$$u_{xx} + \lambda(1 - u) = 0,\qquad(7.3.42)$$

and (iii) the $\lambda = 0$ equation

$$u_t = uu_{xx}.\qquad(7.3.43)$$

The first two equations are ordinary differential equations for which exact finite-difference schemes exist. They are

$$\frac{u^{k+1} - u^k}{\left(\frac{e^{\lambda\Delta t}-1}{\lambda}\right)} = \lambda u^k(1 - u^{k+1}),\qquad(7.3.44)$$

$$\frac{u_{m+1} - 2u_m + u_{m-1}}{\left(\frac{4}{\lambda}\right)\sinh^2\left(\frac{\sqrt{\lambda}\Delta x}{2}\right)} + \lambda(1 - u_m) = 0.\qquad(7.3.45)$$

In the previous section we derived best difference schemes for the $\lambda = 0$ equation, namely, Eqs. (7.3.34) and (7.3.35).

We must now combine Eqs. (7.3.44), (7.3.45) and either Eq. (7.3.34) or (7.3.35) to obtain a discrete model for Eq. (7.3.40). For an explicit scheme there is only one way to do this and the proper scheme is

$$\frac{u_m^{k+1} - u_m^k}{\left(\frac{e^{\lambda\Delta t}-1}{\lambda}\right)} = u_m^{k+1}\left[\frac{u_{m+1}^k - 2u_m^k + u_{m-1}^k}{\left(\frac{4}{\lambda}\right)\sinh^2\left(\frac{\sqrt{\lambda}\Delta x}{2}\right)}\right] + \lambda u_m^k(1 - u_m^{k+1}).\qquad(7.3.46)$$

The corresponding implicit scheme is

$$\frac{u_m^{k+1} - u_m^k}{\left(\frac{e^{\lambda\Delta t}-1}{\lambda}\right)} = u_m^k\left[\frac{u_{m+1}^{k+1} - 2u_m^{k+1} + u_{m-1}^{k+1}}{\left(\frac{4}{\lambda}\right)\sinh^2\left(\frac{\sqrt{\lambda}\Delta x}{2}\right)}\right] + \lambda u_m^k(1 - u_m^{k+1}).\qquad(7.3.47)$$

In contrast to a standard finite-difference model of Eq. (7.3.40), i.e.,

$$\frac{u_m^{k+1} - u_m^k}{\Delta t} = u_m^k\left[\frac{u_{m+1}^k - 2u_m^k + u_{m-1}^k}{(\Delta x)^2}\right] + \lambda u_m^k(1 - u_m^k),\qquad(7.3.48)$$

180

for which we expect a variety of numerical instabilities to occur, the schemes of Eqs. (7.3.46) and (7.3.47) have the following properties:

(i) They have the correct discrete forms for the three special limiting differential equations.

(ii) Nonstandard forms for the discrete derivatives occur.

(iii) The nonlinear terms are modeled by nonlocal discrete expressions on the discrete-space and -time lattice.

7.3.4 $u_t = u_{xx} + \lambda u(1 - u)$

A famous differential equation that was originally used to model mutant-gene propagation is the Fisher equation [25]

$$u_t = u_{xx} + \lambda u(1 - u), \qquad \lambda > 0. \tag{7.3.49}$$

Discrete versions of this equation have been used to investigate numerical instabilities in finite-difference schemes [26, 27].

The Fisher equation has the two sub-equations that are ordinary differential equations. They are

$$u_t = \lambda u(1 - u), \tag{7.3.50}$$

$$u_{xx} + \lambda u(1 - u) = 0. \tag{7.3.51}$$

The exact finite-difference scheme for the first of these equations is given by Eq. (7.3.44). The second differential equation corresponds to a nonlinear conservative oscillator [28]. We now construct a best finite-difference model for it that satisfies an energy conservation principle.

A first integral for Eq. (7.3.51) is [28]

$$\left(\frac{1}{2}\right)u_x^2 + \lambda\left[\frac{u^2}{2} - \frac{u^3}{3}\right] = E = \text{constant}. \tag{7.3.52}$$

The application of standard modeling rules to Eq. (7.3.52) gives

$$\left(\frac{1}{2}\right)\left[\frac{u_m - u_{m-1}}{h}\right]^2 + \lambda\left[\frac{u_m^2}{2} - \frac{u_m^3}{3}\right] = E. \tag{7.3.53}$$

However, this expression does not correspond to a nonlinear oscillator with a conservation law since it is not invariant under the transformation

$$u_m \leftrightarrow u_{m-1}. \tag{7.3.54}$$

(See the discussion presented in Section 5.3.) A nonstandard scheme that does have this property is [29]

$$\left(\frac{1}{2}\right)\left[\frac{u_m - u_{m-1}}{\psi(\Delta t)}\right]^2 + \lambda\left\{\frac{u_m u_{m-1}}{2} - \left(\frac{u_m^2 u_{m-1} + u_m u_{m-1}^2}{6}\right)\right\} = E, \tag{7.3.55}$$

where

$$\psi(h) = h + O(h^2). \tag{7.3.56}$$

Applying the "difference operator," defined as

$$\Delta f_m \equiv f_{m+1} - f_m, \tag{7.3.57}$$

to Eq. (7.3.55) gives the discrete equation of motion

$$\frac{u_{m+1} - 2u_m + u_{m-1}}{[\psi(\Delta t)]^2} + \lambda u_m - \lambda\left(\frac{u_{m+1} + u_m + u_{m-1}}{3}\right)u_m = 0, \tag{7.3.58}$$

where the following results have been used:

$$\Delta(u_m - u_{m-1})^2 = \Delta(u_m^2 - 2u_m u_{m-1} + u_{m-1}^2)$$
$$= (u_{m+1}^2 - u_m^2) - 2(u_{m+1}u_m - u_m u_{m-1}) + (u_m^2 - u_{m-1}^2)$$
$$= (u_{m+1}^2 - u_{m-1}^2) - 2u_m(u_{m+1} - u_{m-1})$$
$$= (u_{m+1} - 2u_m + u_{m-1})(u_{m+1} - u_{m-1}), \tag{7.3.59}$$

$$\Delta u_m u_{m-1} = u_m(u_{m+1} - u_{m-1}), \tag{7.3.60}$$

$$\Delta(u_m^2 u_{m-1} + u_m u_{m-1}^2) = u_m(u_{m+1} + u_m + u_{m-1})(u_{m+1} - u_{m-1}). \qquad (7.3.61)$$

The two discrete sub-equations, Eqs. (7.3.44) and (7.3.58), must now be combined. The only way to do this, to obtain an explicit scheme, is to use the representation

$$\frac{u_m^{k+1} - u_m^k}{\left(\frac{e^{\lambda \Delta t} - 1}{\lambda}\right)} = \frac{u_{m+1}^k - 2u_m^k + u_{m-1}^k}{\left(\frac{4}{\lambda}\right) \sin^2 \left(\frac{\sqrt{\lambda} \Delta x}{2}\right)} + \lambda u_m^k$$
$$- \lambda \left(\frac{u_{m+1}^k + u_m^k + u_{m-1}^k}{3}\right) u_m^{k+1}. \qquad (7.3.62)$$

Corresponding implicit schemes are

$$\frac{u_m^{k+1} - u_m^k}{\left(\frac{e^{\lambda \Delta t} - 1}{\lambda}\right)} = \frac{u_{m+1}^{k+1} - 2u_m^{k+1} + u_{m-1}^{k+1}}{\left(\frac{4}{\lambda}\right) \sin^2 \left(\frac{\sqrt{\lambda} \Delta x}{2}\right)} + \lambda u_m^k$$
$$- \lambda \left(\frac{u_{m+1}^k + u_m^k + u_{m-1}^k}{3}\right) u_m^{k+1}, \qquad (7.3.63)$$

and

$$\frac{u_m^{k+1} - u_m^k}{\left(\frac{e^{\lambda \Delta t} - 1}{\lambda}\right)} = \frac{u_{m+1}^{k+1} - 2u_m^{k+1} + u_{m-1}^{k+1}}{\left(\frac{4}{\lambda}\right) \sin^2 \left(\frac{\sqrt{\lambda} \Delta x}{2}\right)} + \lambda u_m^k$$
$$- \lambda \left(\frac{u_{m+1}^{k+1} + u_m^{k+1} + u_{m-1}^{k+1}}{3}\right) u_m^k. \qquad (7.3.64)$$

Note that the λu term must be evaluated at the m-th discrete-space step and the k-th discrete-time step.

None of the above schemes are even distantly related to the following scheme often used for calculations [26]

$$\frac{u_m^{k+1} - u_m^k}{\Delta t} = \frac{u_{m+1}^k - 2u_m^k + u_{m-1}^k}{(\Delta x)^2} + \lambda u_m^k (1 - u_m^k). \qquad (7.3.65)$$

7.4 Burgers' Type Equations

The Burgers' partial differential equation [2, 3, 30]

$$u_t + u u_x = \nu u_{xx}, \qquad \nu = \text{constant}, \qquad (7.4.1)$$

is a simplification of the Navier-Stokes equations. It is also the governing equation for a variety of one-dimensional flow problems, including, for example, weak shock propagation, compressible turbulence, and continuum traffic simulations [2]. In this section, we present a number of nonstandard discrete models for Burgers' type partial differential equations, i.e., eqnations having the form

$$u_t + uu_x = h_1(u)u_{xx} + h_2(u). \tag{7.4.2}$$

7.4.1 $u_t + uu_x = 0$

The diffusion-free Burgers' equation is

$$u_t + uu_x = 0. \tag{7.4.3}$$

With the initial condition

$$u(x,0) = f(x), \tag{7.4.4}$$

the exact solution is [2]

$$u(x,t) = f[x - u(x,t)t]. \tag{7.4.5}$$

The Burgers' equation has an exact rational solution that can be found by the method of separation of variables. If we write

$$u(x,t) = X(x)T(t), \tag{7.4.6}$$

then $X(x)$ and $T(t)$ satisfy the ordinary differential equations

$$\frac{dT}{dt} = oT^2, \tag{7.4.7}$$

$$\frac{dX}{dx} = -\alpha, \tag{7.4.8}$$

where α is the separation constant. Solving these equations gives

$$T(t) = \frac{1}{A_1 - \alpha t}, \tag{7.4.9}$$

$$X(t) = A_2 - \alpha x, \tag{7.4.10}$$

where A_1 and A_2 are arbitrary constants of integration. Consequently, a special solution of the Burgers' equation is

$$u(x,t) = \frac{A_2 - \alpha x}{A_1 - \alpha t}. \tag{7.4.11}$$

We now require that our finite-difference models for Eq. (7.4.3) have the discrete form of Eq. (7.4.11) as a special exact solution. In addition, we will also impose the condition that the nonlinear term, uu_x, be modeled nonlocally on the discrete-space and -time lattice.

The following two schemes have these properties [19]:

$$\frac{u_m^{k+1} - u_m^k}{\Delta t} + u_m^{k+1}\left(\frac{u_m^k - u_{m-1}^k}{\Delta x}\right) = 0, \tag{7.4.12}$$

$$\frac{u_m^{k+1} - u_m^k}{\Delta t} + u_m^k\left(\frac{u_m^{k+1} - u_{m-1}^{k+1}}{\Delta x}\right) = 0. \tag{7.4.13}$$

Note that applying the method of separation of variables [12] gives for both equations the expressions

$$D_{m+1} - D_m = -\alpha(\Delta x), \tag{7.4.14}$$

$$C^{k+1} - C^k = \alpha(\Delta t)C^{k+1}C^k, \tag{7.4.15}$$

where

$$u_m^k = C^k D_m, \tag{7.4.16}$$

and α is the separation constant. The solutions to these first-order difference equations can be put in the forms

$$D_m = A_2 - \alpha x_m, \tag{7.4.17}$$

$$C^k = \frac{1}{A_1 - \alpha t_k}, \tag{7.4.18}$$

where A_1 and A_2 are arbitrary constants. Therefore,

$$u_m^k = C^k D_m = \frac{A_2 - \alpha x_m}{A_1 - \alpha t_k}, \qquad (7.4.19)$$

and, as stated above, Eqs. (7.4.12) and (7.4.13) both have the same special solutions as the diffusion-free Burgers' equation. Observe that the finite-difference schemes of Eqs. (7.4.12) and (7.4.13) are, respectively, explicit and implicit.

7.4.2 $u_t + uu_x = \lambda u(1 - u)$

The following modified Burgers' equation

$$u_t + uu_x = \lambda u(1 - u), \qquad (7.4.20)$$

where λ is a positive constant, has the three sub-equations

$$u_t = \lambda u(1 - u), \qquad (7.4.21)$$

$$u_x = \lambda(1 - u), \qquad (7.4.22)$$

$$u_t + uu_x = 0. \qquad (7.4.23)$$

The first two equations are ordinary differential equations, while the third is the partial differential equation discussed in the previous section. The exact difference schemes, respectively, for Eqs. (7.4.21) and (7.4.22) are

$$\frac{u^{k+1} - u^k}{\left(\frac{e^{\lambda \Delta t} - 1}{\lambda}\right)} = \lambda u^k(1 - u^{k+1}), \qquad (7.4.24)$$

$$\frac{u_{m+1} - u_m}{\left(\frac{1 - e^{-\lambda \Delta x}}{\lambda}\right)} = \lambda(1 - u_m), \qquad (7.4.25)$$

while best difference schemes for Eq. (7.4.23) are given by Eqs. (7.4.12) and (7.4.13).

We now require that any discrete model of Eq. (7.4.20) reduces, in the appropriate limit, to the finite-difference results given by Eqs. (7.4.24), (7.4.25) and either Eq. (7.4.12) or (7.4.13). There is only one way that this can be done and this scheme is given by the expression

$$\frac{u_m^{k+1} - u_m^k}{\phi(\Delta t)} + u_m^k \left[\frac{u_m^{k+1} - u_{m-1}^{k+1}}{\psi(\Delta x)}\right] = \lambda u_m^k(1 - u_{m-1}^{k+1}), \qquad (7.4.26)$$

where the denominator functions are

$$\phi(\Delta t) = \frac{e^{\lambda \Delta t} - 1}{\lambda}, \qquad \psi(\Delta x) = \frac{1 - e^{-\lambda \Delta x}}{\lambda}. \qquad (7.4.27)$$

This finite-difference scheme has the following properties:

(i) The discrete model is implicit.

(ii) The denominator functions are not of the simple forms $\phi(\Delta t) = \Delta t$ and $\psi(\Delta x) = \Delta x$.

(iii) The nonlinear terms are modeled nonlocally on the discrete-space and -time lattice.

(iv) The discrete space-derivative is a backward Euler type scheme.

7.4.3 $u_t + uu_x = uu_{xx}$

Consider the following modified Burgers' equation [32]

$$u_t + uu_x = uu_{xx}. \qquad (7.4.28)$$

This nonlinear partial differential equation does not have a known exact general solution that can be written in terms of a finite number of elementary functions. However, a special solution can be found by use of the method of separation of variables. Assuming for $u(x,t)$ the form

$$u(x,t) = X(x)T(t), \qquad (7.4.29)$$

we find that

$$u(x,t) = \frac{A + Be^x + Cx}{Ct + D}, \qquad (7.4.30)$$

where (A, B, C, D) are arbitrary constants. Therefore, we require that our finite-difference schemes also have the discrete version of Eq. (7.4.30) as a special solution.

The application of the standard rules to Eq. (7.4.28) gives, for example, the form

$$\frac{u_m^{k+1} - u_m^k}{\Delta t} + u_m^k \left(\frac{u_{m+1}^k - u_{m-1}^k}{2\Delta x} \right) = u_m^k \left[\frac{u_{m+1}^k - 2u_m^k + u_{m-1}^k}{(\Delta x)^2} \right]. \qquad (7.4.31)$$

The method of variables can be applied to this difference equation. Writing u_m^k as

$$u_m^k = X_m T^k, \qquad (7.4.32)$$

we find that X_m and T^k satisfy the equations

$$\frac{X_{m+1} - 2X_m + X_{m-1}}{(\Delta x)^2} - \left[\frac{X_{m+1} - X_{m-1}}{2\Delta x} \right] = -C, \qquad (7.4.33)$$

$$\frac{T^{k+1} - T^k}{\Delta t} = -C(\Delta t)(T^k)^2, \qquad (7.4.34)$$

where C is the separation constant. The solutions to these equations do not correspond to the discrete versions of $X(x)$ and $T(t)$ as given in Eq. (7.4.30).

Now consider the following nonstandard model for Eq. (7.4.28) [32]:

$$\frac{u_m^{k+1} - u_m^k}{\Delta t} + u_m^{k+1} \left(\frac{u_m^k - u_{m-1}^k}{\Delta x} \right) = u_m^{k+1} \left[\frac{u_{m+1}^k - 2u_m^k + u_{m-1}^k}{(e^{\Delta x} - 1)\Delta x} \right]. \qquad (7.4.35)$$

This result is obtained by combining the finite-difference schemes for the three sub-equations

$$u_t + u u_x = 0, \qquad (7.4.36)$$

$$u_t = u u_{xx}, \qquad (7.4.37)$$

$$u_x = u_{xx}. \qquad (7.4.38)$$

Best schemes for Eqs. (7.4.36) and (7.4.37) have already been given, see Eqs. (7.4.12) and (7.3.34). The exact scheme for Eq. (7.4.38) is [19]

$$\frac{u_m - u_{m-1}}{\Delta x} = \frac{u_{m+1} - 2u_m + u_{m-1}}{(e^{\Delta x} - 1)\Delta x}. \qquad (7.4.39)$$

A detailed examination of Eq. (7.4.35) leads to the following conclusions:

(i) This nonstandard scheme has an exact solution that can be found from the method of separation of variables. It is given by the discrete form of Eq. (7.4.30), namely,

$$u_m^k = \frac{A + Be^{x_m} + Cx_m}{Ct_k + D}.$$

(7.4.40)

See reference [32].

(ii) All the nonlinear terms of Eq. (7.4.28) are modeled nonlocally.

(iii) The denominator function for the discrete second-derivative has a nonstandard form.

(iv) The first-derivative is given by a backward Euler type expression.

(v) The finite-difference scheme is explicit.

(vi) A fully implicit scheme is given by the expression

$$\frac{u_m^{k+1} - u_m^k}{\Delta t} + u_m^k \left(\frac{u_m^{k+1} - u_{m-1}^{k+1}}{\Delta x} \right) = u_m^k \left[\frac{u_{m+1}^{k+1} - 2u_m^{k+1} + u_{m-1}^{k+1}}{(e^{\Delta x} - 1)\Delta x} \right].$$

(7.4.41)

7.5 Discussion

In general, we do not expect exact finite-difference schemes to exist for arbitrary partial differential equations [18, 9]. However, in practical applications, best schemes can be found and they should provide better discrete models than those obtained from use of standard methods [8, 10, 24, 29, 31, 32]. However, the work of this chapter shows that for a given partial differential equation, more than one best scheme may exist. This non-uniqueness usually appears in the form of the existence of both explicit and implicit best schemes for the equation of interest. Resolution of this problem can only come from imposing additional requirements on the finite-difference schemes.

The "toy" partial differential equations considered in this chapter were investigated because they have special solutions that can be found and/or they have subequations such that these equations can be solved exactly or have special solutions that can be discovered. Our basic procedure for constructing best finite-difference schemes consisted of imposing one or both of the following requirements on the discrete model:

(i) Special solutions of the differential equation should also be special solutions of the finite-difference equation.

(ii) Corresponding sub-equations of the differential and finite-difference equations should have (essentially) the same mathematical properties. In particular, this means that in the proper limits, the sub-equations of the discrete equation should reduce to the correct differential equations.

The key to success in the formulation and construction of best finite-difference schemes is the application of the nonstandard modeling rules as given in Section 3.4. Also, it is equally important to be knowledgeable of both exact and best finite-difference schemes for a large number of ordinary and partial differential equations.

For those few partial differential equations for which the general solution can be found, it is always the case that a functional relation exists between the space and time step-sizes, i.e.,

$$\Delta t = \Sigma(\Delta x). \tag{7.5.1}$$

Unfortunately, most of the best schemes constructed in this chapter do not allow the determination of such a relation. Additional information is required to obtain such restrictions. The imposition of conditional stability often provides this functional relation for linear equations.

The partial differential equations investigated in this chapter have only quadratic nonlinearities. It would be of great value to generalize these procedures to other types of nonlinear terms and other classes of equations.

References

1. W. F. Ames, *Nonlinear Partial Differential Equations in Engineering* (Academic Press, New York, 1965).

2. G. B. Whitham, *Linear and Nonlinear Waves* (Wiley-Interscience, New York, 1974).

3. D. Potter, *Computational Physics* (Wiley-Interscience, New York, 1973).

4. F. B. Hildebrand, *Finite-Difference Equations and Simulations* (Prentice-Hall; Englewood Cliffs, NJ; 1968).

5. R. D. Richtmyer and K. W. Morton, *Difference Methods for Initial-Value Problems* (Interscience, New York, 2nd edition, 1967).

6. G. D. Smith, *Numerical Solution of Partial Differential Equations: Finite Difference Methods* (Clarendon Press, Oxford, 1978).

7. A. R. Mitchell and D. F. Griffiths, *Finite Difference Methods in Partial Differential Equations* (Wiley, New York, 1980).

8. R. E. Mickens, *Journal of Sound and Vibration* 100, 452–455 (1985). Exact finite difference schemes for the nonlinear unidirectional wave equation.

9. R. E. Mickens, *Numerical Methods for Partial Differential Equations* 5, 313–325 (1989). Exact solutions to a finite-difference model of a nonlinear reaction-advection equation: Implications for numerical analysis.

10. R. E. Mickens, *Journal of Sound and Vibration* 146, 342–344 (1991). Analysis of a new finite-difference scheme for the linear advection-diffusion equation.

11. E. Zauderer, *Partial Differential Equations of Applied Mathematics* (Wiley-Interscience, New York, 1983).

12. R. E. Mickens, *Difference Equations: Theory and Applications* (Van Nostrand Reinhold, New York, 2nd edition, 1990).

13. R. Peyret and T. D. Taylor, *Computational Methods for Fluid Flow* (Springer-Verlag, New York, 1983).

14. T. F. Chan, *SIAM Journal of Numerical Analysis* 21, 272–284 (1984). Stability analysis of finite-difference schemes for the advection-diffusion equation.

15. J. C. Strikwerda, *Finite Difference Schemes and Partial Differential Equations* (Wadsworth; Pacific Grove, CA; 1989).

16. L. R. Bentley, G. F. Pinder and I. Herrera, *Numerical Methods for Partial Differential Equations* **5**, 227–240 (1989). Solution of the advective-dispersive transport equation using a least squares collocation, Eulerian-Lagrangian method.

17. G. E. Forsythe and W. R. Wasow, *Finite-Difference Methods for Partial Differential Equations* (Wiley, New York, 1960).

18. R. E. Mickens, Pitfalls in the numerical integration of differential equations, in *Analytical Techniques for Material Characterization*, W. E. Collins et al., editors (World Scientific Publishing, Singapore, 1987), pp. 123–143.

19. R. E. Mickens, *Numerical Methods for Partial Differential Equations* **2**, 123–129 (1986). Exact solutions to difference equation models of Burgers' equation.

20. D. Greenspan and V. Casulli, *Numerical Analysis for Applied Mathematics, Science and Engineering* (Addison-Wesley; Redwood City, CA; 1988).

21. H. S. Carslaw and J. C. Jaeger, *Conduction of Heat in Solids* (Clarendon, London, 2nd edition, 1959).

22. J. Crank and P. Nicolson, *Proceedings of the Cambridge Philosophical Society* **43**, 50–67 (1947). A practical method for numerical evaluation of solutions of partial differential equations of heat conduction type.

23. M. C. Bhattacharya, *Communications in Applied Numerical Methods* **6**, 173–184 (1990). Finite-difference solutions of partial differential equations.

24. R. E. Mickens, *Numerical Methods for Partial Differential Equations* **7**, 299–302 (1991). Construction of a novel finite-difference scheme for a nonlinear diffusion equation.

25. R. A. Fisher, *Annuals of Eugenics* **7**, 355 (1937). The wave of advance of advantageous genes.

26. A. R. Mitchell and J. C. Bruch, Jr., *Numerical Methods for Partial Differential Equations* **1**, 13–23 (1985). A numerical study of chaos in a reaction-diffusion equation.

27. N. Parekh and S. Puri, *Physical Review E* **2**, 1415–1418 (1993). Velocity selection in coupled-map lattices.

28. J. B. Marion, *Classical Dynamics of Particles and Systems* (Academic Press, New York, 2nd edition, 1970).

29. R. E. Mickens, "New finite-difference scheme for the Fisher equation," Clark Atlanta University preprint; June 1993.

30. J. M. Burgers, *Advances in Applied Mechanics* **1**, 171–199 (1948). A mathematical model illustrating the theory of turbulence.

31. R. E. Mickens and J. N. Shoosmith, *Journal of Sound and Vibration* **142**, 536–539 (1990). A discrete model of a modified Burgers' partial differential equation.

32. R. E. Mickens, *Transactions of the Society for Computer Simulation* **8**, 109–117 (1991). Nonstandard finite difference schemes for partial differential equations.

Chapter 8
SCHRÖDINGER DIFFERENTIAL EQUATIONS

8.1 Introduction

Schrödinger type ordinary and partial differential equations arise in the modeling of a large number of physical phenomena. Particular areas include quantum mechanics, ocean acoustics, optics, plasma physics and seismology [1–4]. A large literature exists on the determination of asymptotic techniques for calculating analytic approximations to the solutions of these equations [1, 5, 6]. The work presented in this chapter will center on the construction of finite-difference techniques for use in the numerical integration of Schrödinger type ordinary and partial differential equations in one space dimension [7–9].

For our purposes, the Schrödinger partial differential equation takes the form

$$\frac{\partial u}{i\partial t} = \frac{\partial^2 u}{\partial x^2} + f(x)u. \tag{8.1.1}$$

This equation is usually called the time-dependent Schrödinger equation. The related Schrödinger time-independent ordinary differential equation can be expressed as

$$\frac{d^2 u}{dx^2} + \bar{f}(x)u = 0, \tag{8.1.2}$$

where, for many applications, $f(x)$ and $\bar{f}(x)$ differ only by a constant [1].

In Section 8.2, we discuss a novel finite-difference scheme for Eq. (8.1.2) [9]. The generalization of this procedure to include the Numerov scheme [10] is then presented [11]. Section 8.3 begins with an examination of the difficulties of constructing stable finite-difference schemes for the free-particle Schrödinger equation

$$\frac{\partial u}{i\partial t} = \frac{\partial^2 u}{\partial x^2}. \tag{8.1.3}$$

We show that this problem can, in part, be overcome by using the concept of nonstandard discrete derivatives [7, 8]. Next, this result is used to construct a novel

discrete model for the time-dependent Schrödinger equation. We end the section
with an application of these ideas to the nonlinear, cubic Schrödinger equation.

8.2 Schrödinger Ordinary Differential Equations

8.2.1 Numerov Method

The simplest finite-difference scheme for the time-independent Schrödinger
equation

$$\frac{d^2u}{dx^2} + f(x)u = 0, \tag{8.2.1}$$

(where we have dropped the bar over the function $f(x)$, see Eq. (8.1.2)) is

$$\frac{u_{m+1} - 2u_m + u_{m-1}}{h^2} + f_m u_m = 0, \tag{8.2.2}$$

where

$$f_m = f(x_m), \qquad x_m = (\Delta x)m = hm. \tag{8.2.3}$$

However, for a variety of reasons, including the possible existence of numerical
instabilities and the lack of numerical accuracy, other discrete models have been
considered [10, 12, 13]. A popular method is the Numerov algorithm [10]. The
starting point for deriving this scheme is the relation [14]

$$\frac{u_{m+1} - 2u_m + u_{m-1}}{h^2} = u_m'' + \left(\frac{h^2}{12}\right)u_m'''' + O(h^4), \tag{8.2.4}$$

where

$$u_m = u(x_m), \qquad f_m = f(x_m), \tag{8.2.5}$$

and

$$u_m'' = \left.\frac{d^2u}{dx^2}\right|_{x=x_m}, \qquad u_m'''' = \left.\frac{d^4u}{dx^4}\right|_{x=x_m} \tag{8.2.6}$$

From Eq. (8.2.1), we have

$$u_m'' = -f_m u_m. \tag{8.2.7}$$

Consequently,

$$u_m'''' = -\frac{d^2}{dx^2} [f(x)u]\Big|_{x=x_m}$$

$$= -\left[\frac{f_{m+1}u_{m+1} - 2f_m u_m + f_{m-1}u_{m-1}}{h^2}\right] + O(h^2). \qquad (8.2.8)$$

Substitution of Eqs. (8.2.7) and (8.2.8) into Eq. (8.2.4) gives

$$\frac{u_{m+1} - 2u_m + u_{m-1}}{h^2} = -f_m u_m$$

$$- \left(\frac{h^2}{12}\right)\left[\frac{f_{m+1}u_{m+1} - 2f_m u_m + f_{m-1}u_{m-1}}{h^2}\right]$$

$$+ O(h^4). \qquad (8.2.9)$$

Simplifying this expression and neglecting terms of $O(h^4)$ gives the Numerov algorithm

$$\left[1 + \frac{h^2 f_{m+1}}{12}\right] u_{m+1} - 2\left[1 - \frac{5h^2 f_m}{12}\right] u_m + \left[1 + \frac{h^2 f_{m-1}}{12}\right] u_{m-1} = 0. \qquad (8.2.10)$$

8.2.2 Mickens-Ramadhani Scheme

This finite-difference scheme proposed for the time-independent Schrödinger equation is based on the use of nonstandard modeling rules for the construction of discrete representations for differential equations; see Section 3.4 and the references [15, 16]. We begin with the exact difference scheme for

$$\frac{d^2u}{dx^2} + \lambda u = 0 \qquad (8.2.11)$$

where λ is a constant. It is given by the expression

$$\frac{u_{m+1} - 2u_m + u_{m-1}}{\left(\frac{4}{\lambda}\right)\sin^2\left(\frac{h\sqrt{\lambda}}{2}\right)} + \lambda u_m = 0. \qquad (8.2.12)$$

Note that this relation holds whether λ is positive or negative. This result follows directly from the use of the relation:

$$\sin(i\theta) = i\sinh(\theta). \qquad (8.2.13)$$

The Mickens-Ramadhani scheme replaces the constant λ is Eq. (8.2.12) by the discrete form of the function $f(x)$, i.e.,

$$\lambda \rightarrow f_m = f(x_m). \tag{8.2.14}$$

This gives the scheme

$$\frac{u_{m+1} - 2u_m + u_{m-1}}{\left(\frac{4}{f_m}\right) \sin^2 \left(\frac{h\sqrt{f_m}}{2}\right)} + f_m u_m = 0. \tag{8.2.15}$$

Using the trigonometric identity

$$2\sin^2 \theta = 1 - \cos 2\theta, \tag{8.2.16}$$

we obtain

$$u_{m+1} + u_{m-1} = 2\left[\cos(h\sqrt{f_m})\right] u_m. \tag{8.2.17}$$

For purposes of comparison, we rewrite the simple finite-difference scheme of Eq. (8.2.2) in this form; it is

$$u_{m+1} + u_{m-1} = 2\left(1 - \frac{h^2 f_m}{2}\right) u_m. \tag{8.2.18}$$

Asymptotic behavior of the solutions to difference equations can vary widely between two equations that seemingly have minor differences in their structure. Thus, it is of interest to compare the solutions of the Mickens-Ramadhani and standard schemes to that of the discrete version of a differential equation with known asymptotic solution. We select the zero-th order Bessel equation

$$\frac{d^2 w}{dx^2} + \left(\frac{1}{x}\right) \frac{dw}{dx} + w = 0. \tag{8.2.19}$$

The transformation [5]

$$\sqrt{x}\, w = u, \tag{8.2.20}$$

converts Eq. (8.2.19) to the equation

$$\frac{d^2u}{dx^2} + \left(1 + \frac{1}{4x^2}\right)u = 0. \qquad (8.2.21)$$

Using the WKB procedure, the following asymptotic $(x \to \infty)$ solution is obtained [5]

$$u(x) = A\left[\sin(x) - \frac{\cos(x)}{8x}\right] + B\left[\cos(x) + \frac{\sin(x)}{8x}\right] + O\left(\frac{1}{x^2}\right), \qquad (8.2.22)$$

where A and B are arbitrary constants.

Discrete versions of the WKB method also exist for calculating the asymptotic solutions of linear second-order difference equations [6, 17, 18]. Applying these procedures to Eqs. (8.2.17) and (8.2.18) gives, respectively [9], for $f_m = [1 + (4x_m^2)^{-1}]$,

$$u_m^{(MR)} = A\left[\sin(x_m) - \frac{\cos(x_m)}{8x_m}\right] + B\left[\cos(x_m) + \frac{\sin(x_m)}{8x_m}\right] + O\left(\frac{1}{x_m^2}\right), \qquad (8.2.23)$$

$$\begin{aligned} u_m^{(S)} = A&\left\{\sin[\phi(h)x_m] - \frac{\beta(h)\cos[\phi(h)x_m]}{8x_m}\right\} \\ + B&\left\{\cos[\phi(h)x_m] + \frac{\beta(h)\sin[\phi(h)x_m]}{8x_m}\right\} \\ + O&\left(\frac{1}{x_m^2}\right), \end{aligned} \qquad (8.2.24)$$

where

$$\phi(h) = \left(\frac{1}{h}\right)\tan^{-1}\left[\frac{\sqrt{4h^2 - h^4}}{2 - h^2}\right], \qquad (8.2.25)$$

$$\beta(h) = h^2\left[1 - \frac{h^2}{4}\right]^{1/2} + \frac{(1 - h^2/2)^2}{\sqrt{1 - h^2/4}}. \qquad (8.2.26)$$

Note that the Mickens-Ramadhani scheme agrees with the exact result to terms of $O(x_m^{-2})$, while the standard scheme always disagrees with the exact answer for finite step-size h.

8.2.3 Combined Numerov-Mickens Scheme

A finite-difference scheme for the time-independent Schrödinger equation that combines the Numerov and Mickens-Ramadhani schemes has been constructed and studied by Chen et al. [11]. This scheme has the form

$$
\left[1 + \frac{h^2 f_{m+1}}{12}\right] u_{m+1} + \left[1 + \frac{h^2 f_{m-1}}{12}\right] u_{m-1}
$$
$$
= 2\left[\cos\left(h\sqrt{f_m}\right)\right]\left[1 + \frac{h^2 f_m}{12}\right] u_m. \qquad (8.2.27)
$$

They call the new discrete model the "combined Numerov-Mickens finite-difference scheme" (CNMFDS). Examination of the Numerov, Mickens-Ramadhani and CNMFDS representations allows the following conclusions to be reached:

(i) The Mickens-Ramadhani scheme is (formally) of $O(h^2)$, while the Numerov scheme is $O(h^4)$.

(ii) The Mickens-Ramadhani scheme is an exact finite-difference model for $f(x) = $ constant. This is not the situation for the Numerov scheme.

(iii) The CNMFDS is of $O(h^4)$, just like the Numerov scheme, and it is also an exact finite-difference scheme for $f(x) = $ constant.

Numerical experiments were also done by Chen et al. [11]. Their general conclusion was that the CNMFDS has potential for use in practical calculations.

8.3 Schrödinger Partial Differential Equations

Finite-difference schemes for Schrödinger partial differential equations generally separate into two classes: implicit and explicit formulations [19–22]. Most investigations have focused on implicit schemes because of the good stability properties that these schemes possess. (Stability is used here in the sense that small errors in the initial data do not grow as the discrete-time is increased [23, 24, 25].) However, a major difficulty with implicit schemes is the need to solve large sets of systems of complex-valued algebraic equations. In contrast, many explicit schemes, such as a simple forward Euler scheme, are unconditionally unstable [19, 20]. However, explicit schemes are generally easier to implement and have fewer computer-storage requirements as compared to implicit schemes.

In Section 8.3.1, we show that it is possible to construct explicit, forward Euler schemes for the so-called free-particle Schrödinger equation. This construction is based on the use of a nonstandard denominator function for the discrete-time derivative. These schemes are conditionally stable. In Section 8.3.2, we apply these results and the use of nonstandard modeling rules to obtain finite-difference models for the full time-dependent Schrödinger equation. Finally, in Section 8.3.3, an application is made of these procedures to the nonlinear, cubic Schrödinger equation.

8.3.1 $u_t = iu_{xx}$

The simplest Schrödinger type partial differential equation is the free-particle equation [1, 19, 20]

$$\frac{\partial u}{i\partial t} = \frac{\partial^2 u}{\partial x^2}.$$ (8.3.1)

The direct forward Euler scheme

$$\frac{u_m^{k+1} - u_m^k}{i\Delta t} = \frac{u_{m+1}^k - 2u_m^k + u_{m-1}^k}{(\Delta x)^2},$$ (8.3.2)

is unconditionally unstable for any choice of Δx and Δt [19, 20]. We now demonstrate that a conditionally stable finite-difference model can be constructed using nonstandard modeling rules [7].

The following is a list of properties that an explicit finite-difference model for Eq. (8.3.1) should possess:

(i) The discrete-time derivative must be of first-order.

(ii) The discrete-space derivative must be of second-order and centered about x_m.

(iii) There should not be any *ad hoc* terms in the scheme.

(iv) The finite-difference scheme should be (at least) conditionally stable.

Requirements (i) and (ii) are consequences of Eq. (8.3.1) being a partial differential equation that is first-order in the time derivative and second-order in the space derivative. A higher order scheme for either discrete derivative would lead to numerical instabilities [15, 16]. Condition (iii) is the requirement that the finite-difference scheme should be "natural," i.e., very term in the discrete model should have a counterpart in the differential equation. Finally, requirement (iv) is needed to ensure that practical calculations can actually be done using the scheme.

It should be indicated that these requirements automatically eliminate from consideration several discrete schemes for the Schrödinger partial differential equation. One example is the explicit finite-difference scheme of Cahn et al. [19],

$$\frac{u_m^{k+1} - u_m^k}{i\Delta t} = \frac{u_{m+1}^k - 2u_m^k + u_{m-1}^k}{(\Delta x)^2}$$
$$+ (\alpha + i\beta)\Delta t \left[\frac{u_{m+2}^k - 4u_{m+1}^k + 6u_m^k - 4u_{m-1}^k + u_{m-2}^k}{(\Delta x)^2}\right], \qquad (8.3.3)$$

where α and β are certain constants. This form is eliminated because of the *ad hoc* nature of the second term on the right-side of the equation. The central difference discrete-time form [26]

$$\frac{u_m^{k+1} - u_m^{k-1}}{2i\Delta t} = \frac{u_{m+1}^k - 2u_m^k + u_{m-1}^k}{(\Delta x)^2}, \qquad (8.3.4)$$

is also eliminated since the partial difference equation is second-order in the discrete-time variable and thus violates condition (i).

Consider now the first-order ordinary differential equation

$$\frac{dw}{dt} = i\lambda w, \qquad (8.3.5)$$

where λ is an arbitrary real number. The exact finite-difference scheme for it is

$$\frac{w_{k+1} - w_k}{\left(\frac{e^{i\lambda h} - 1}{i\lambda}\right)} = i\lambda w_k, \qquad h = \Delta t. \qquad (8.3.6)$$

A conventional discrete model for Eq. (8.3.5) is

$$\frac{w_{k+1} - w_k}{h} = i\lambda w_k. \tag{8.3.7}$$

Note that the denominator function, $D(h, \lambda)$, for the exact scheme is

$$D(h, \lambda) = \frac{e^{i\lambda h} - 1}{i\lambda} = h + i\left(\frac{\lambda h^2}{2}\right) + O(\lambda^2 h^3), \tag{8.3.8}$$

in contrast to the simple denominator function, h, for Eq. (8.3.7). This result suggests the following form for an explicit scheme for Eq. (8.3.1)

$$\frac{u_m^{k+1} - u_m^k}{i D_1(\Delta t, \lambda)} = \frac{u_{m+1}^k - 2u_m^k + u_{m-1}^k}{D_2(\Delta x, \lambda)}, \tag{8.3.9}$$

where the denominator functions have the properties

$$D_1(\Delta t, \lambda) = \Delta t + i\lambda(\Delta t)^2 + O[(\Delta t)^3], \tag{8.3.10}$$

$$D_2(\Delta x, \lambda) = (\Delta x)^2 + O[(\Delta x)^4]. \tag{8.3.11}$$

At this stage of the analysis, the exact dependencies of D_1 and D_2 on $(\Delta x, \Delta t, \lambda)$ do not have to be specified in any more detail than that given by the conditions of Eqs. (8.3.10) and (8.3.11). These requirements ensure that the finite-difference scheme is both convergent to and consistent with the original partial differential equation.

Define $R(\Delta t, \Delta x, \lambda)$ to be

$$R \equiv \frac{i D_1(\Delta t, \lambda)}{D_2(\Delta x, \lambda)}. \tag{8.3.12}$$

With this definition Eq. (8.3.9) can be rewritten to the form

$$u_m^{k+1} = R u_{m+1}^k + (1 - 2R)u_m^k + R u_{m-1}^k. \tag{8.3.13}$$

We now need to do a stability analysis for this finite-difference scheme. The stability concept to be applied is the von Neumann or Fourier-series method [23–25]. This procedure is based on the fact that both Eqs. (8.3.1) and (8.3.9) are linear equations and the physical solutions of Eq. (8.3.1) are bounded for all times. With this in mind, the stability properties of Eq. (8.3.13) can be studied by considering a typical Fourier mode

$$u_m^k = C(k)e^{i\omega(\Delta x)m},$$

(8.3.14)

where ω is a constant, and requiring that $C(k)$ be bounded for all k. This concept of stability is called practical stability [19, 23, 24].

The substitution of Eq. (8.3.14) into Eq. (8.3.13) gives the following equation for $C(k)$:

$$C(k+1) = AC(k)$$

(8.3.15)

where

$$A = 1 - 4R\sin^2\left(\frac{\theta}{2}\right), \qquad \theta = \omega(\Delta x).$$

(8.3.16)

The solution to this first-order difference equation is

$$C(k) = C(0)A^k,$$

(8.3.17)

where $C(0)$ is an arbitrary constant. Now, if $C(k)$ is to be bounded, then A must satisfy the condition

$$|A| \leq 1.$$

(8.3.18)

Let R, see Eq. (8.3.12), be written as

$$R = R_1 + iR_2,$$

(8.3.19)

where R_1 and R_2 are real functions of Δt, Δx and λ. A straightforward calculation shows that the inequality of Eq. (8.3.18) is equivalent to the expression

$$R_1^2 + R_2^2 \leq \frac{R_1}{2},$$

(8.3.20)

or

$$\left(R_1 - \frac{1}{4}\right)^2 + R_2^2 \leq \frac{1}{16}.$$

(8.3.21)

This inequality has the following geometric interpretation: In the (R_1, R_2) plane, the finite-difference scheme of Eq. (8.3.13) is stable for all points on and inside the semi-circle of radius 0.25, centered at $(0.25, 0)$, and lying in the upper plane. We will refer to this relation as the *circle condition* [7, 27].

In practice, the circle condition is to be used as follows:

(a) Select denominator functions with the properties given by Eqs. (8.3.10) and (8.3.11).

(b) Calculate $R(\Delta t, \Delta x, \lambda)$ as given by Eq. (8.3.12).

(c) Calculate R_1 and R_2, respectively, the real and imaginary parts of R.

(d) Select a point $(\bar{R}_1, \bar{R}_2)$ consistent with the circle condition of Eq. (8.3.21).

(e) Set $R_1(\Delta t, \Delta x, \lambda)$ and $R_2(\Delta t, \Delta x, \lambda)$ equal, respectively, to $\bar{R}_1$ and $\bar{R}_2$, i.e.,

$$R_1(\Delta t, \Delta x, \lambda) = \bar{R}_1, \qquad R_2(\Delta t, \Delta x, \lambda) = \bar{R}_2. \qquad (8.3.22)$$

(f) Choose a value for the space step-size, Δx, and solve Eqs. (8.3.22) for Δt and λ in terms of Δx. Carrying out these operations gives

$$\Delta t = f_1(\Delta x), \qquad \lambda = f_2(\Delta x). \qquad (8.3.23)$$

Therefore, the selection of the point $(\bar{R}_1, \bar{R}_2)$, satisfying the circle condition and the relations of Eq. (8.3.23), completely defines the explicit finite-difference scheme given by Eq. (8.3.9) or (8.3.13).

The Schrödinger equation, in general, has solutions for which the "amplitude" does not decrease with the increase of time. In fact, for physical applications, the solutions to Eq. (8.3.1) satisfy the requirement [1]

$$\int_{-\infty}^{\infty} |u(x,t)|^2 dx = 1. \qquad (8.3.24)$$

Therefore, the circle condition of Eq. (8.3.21) should read

$$\left(R_1 - \frac{1}{4}\right)^2 + R_2^2 = \frac{1}{16}. \qquad (8.3.25)$$

Let us apply this result to the simple finite-difference scheme of Eq. (8.3.2). For this case, we have

$$D_1 = \Delta t, \qquad D_2 = (\Delta x)^2, \tag{8.3.26}$$

$$R = \frac{i\Delta t}{(\Delta x)^2}, \tag{8.3.27}$$

$$R_1 = 0, \qquad R_2 = \frac{\Delta t}{(\Delta x)^2}. \tag{8.3.28}$$

Substitution of these values into the circle condition of Eq. (8.3.21) gives

$$R_2^2 = 0. \tag{8.3.29}$$

This cannot be satisfied for any finite values of Δt and Δx. Consequently, we conclude that the simple forward Euler scheme is unstable. This is the same conclusion that is reached by the usual methods of stability analysis [19, 20].

Let us now consider a finite-difference scheme for Eq. (8.3.1) such that [7, 8]

$$D_1 = \Delta t + i\lambda(\Delta t)^2, \qquad D_2 = (\Delta t)^2. \tag{8.3.30}$$

This corresponds to the use of a nonstandard discrete time-derivative and a standard discrete space-derivative. Therefore,

$$R = -\lambda\left(\frac{\Delta t}{\Delta x}\right)^2 + i\left[\frac{\Delta t}{(\Delta x)^2}\right], \tag{8.3.31}$$

with

$$R_1 = -\lambda\left(\frac{\Delta t}{\Delta x}\right)^2, \qquad R_2 = \frac{\Delta t}{(\Delta x)^2}. \tag{8.3.32}$$

Selection of the (circle condition) point $(\bar{R}_1, \bar{R}_2)$

$$\bar{R}_1 = \frac{1}{4}, \qquad \bar{R}_2 = \frac{1}{4}, \tag{8.3.33}$$

gives

$$-\lambda\left(\frac{\Delta t}{\Delta x}\right)^2 = \frac{1}{4}, \qquad \frac{\Delta t}{(\Delta x)^2} = \frac{1}{4}. \tag{8.3.34}$$

These equations can be solved for Δt and λ in terms of Δx. Doing this gives

$$\Delta t = \frac{(\Delta x)^2}{4}, \qquad \lambda = -\frac{4}{(\Delta x)^2}, \tag{8.3.35}$$

and

$$D_1 = \left(\frac{1-i}{4}\right)(\Delta x)^2, \qquad R = \frac{1+i}{4}. \tag{8.3.36}$$

Consequently, our nonstandard explicit finite-difference scheme for the free-particle Schrödinger partial differential equation is [7]

$$\frac{u_m^{k+1} - u_m^k}{i\left(\frac{1-i}{4}\right)(\Delta x)^2} = \frac{u_{m+1}^k - 2u_m^k + u_{m-1}^k}{(\Delta x)^2}, \tag{8.3.37}$$

or

$$\frac{u_m^{k+1} - u_m^k}{i\left(\frac{1+i}{4}\right)} = u_{m+1}^k - 2u_m^k + u_{m-1}^k, \tag{8.3.38}$$

with

$$\Delta t = \frac{(\Delta x)^2}{4}. \tag{8.3.39}$$

The above scheme is called conditionally stable since practical stability [19, 23] holds only if the functional relation of Eq. (8.3.39) is satisfied for the step-sizes.

For completeness, we now investigate the stability properties of two other standard finite-difference models for Eq. (8.3.1). The use of a central-difference discrete time-derivative gives the scheme

$$\frac{u_m^{k+1} - u_m^{k-1}}{2i\Delta t} = \frac{u_{m+1}^k - 2u_m^k + u_{m-1}^k}{(\Delta x)^2}. \tag{8.3.40}$$

Let β be defined as

$$\beta = \frac{2i\Delta t}{(\Delta x)^2} = i\bar{\beta}. \tag{8.3.41}$$

Using this, Eq. (8.3.40) can be written as

$$u_m^{k+1} = u_m^{k-1} + i\bar{\beta}(u_{m+1}^k - 2u_m^k + u_{m-1}^k). \tag{8.3.42}$$

Substituting

$$u_m^k = C(k)e^{i\omega(\Delta x)m} = C(k)e^{i\theta m} \tag{8.3.43}$$

into Eq. (8.3.42) gives the following linear, second-order difference equation for $C(k)$:

$$C(k+1) + [2i\bar{\beta}(1 - \cos\theta)]C(k) - C(k-1) = 0. \qquad (8.3.44)$$

The solution to this equation has the form [28]

$$C(k) = B_1(r_1)^k + B_2(r_2)^k, \qquad (8.3.45)$$

where r_1 and r_2 are solutions to the characteristic equation

$$r^2 + [2i\bar{\beta}(1 - \cos\theta)]r - 1 = 0. \qquad (8.3.46)$$

Therefore,

$$r_{1,2} = -i\bar{\beta}(1 - \cos\theta) \pm \sqrt{1 - \bar{\beta}^2(1 - \cos\theta)^2}, \qquad (8.3.47)$$

and

$$|r_1| = 1, \qquad |r_2| = 1. \qquad (8.3.48)$$

These results imply that $C(k)$ oscillates with a constant amplitude. Hence, the scheme given by Eq. (8.3.40) is unconditionally stable. In other words, this scheme has practical stability for all values of Δx and Δt. Note, however, that this discrete model is eliminated by our nonstandard modeling rules since the discrete time-derivative is of second-order.

Now consider a standard implicit finite-difference scheme. This is given by the expression

$$\frac{u_m^{k+1} - u_m^k}{i\Delta t} = \frac{u_{m+1}^{k+1} - 2u_m^{k+1} + u_{m-1}^{k+1}}{(\Delta x)^2}. \qquad (8.3.49)$$

Defining β_1 to be

$$\beta_1 = \frac{i\Delta t}{(\Delta x)^2}, \qquad (8.3.50)$$

we have

$$u_m^{k+1} = \beta_1(u_{m+1}^{k+1} + u_{m-1}^{k+1}) - 2\beta_1 u_m^{k+1} + u_m^k. \qquad (8.3.51)$$

Substituting u_m^k from Eq. (8.3.43) into this equation gives the following linear first-order difference equation for $C(k)$:

$$[1 + 2\beta_1(1 - \cos\theta)]C(k+1) = C(k). \qquad (8.3.52)$$

Since

$$|[1 + 2\beta_1(1 - \cos\theta)]| \geq 1, \qquad (8.3.53)$$

it can be concluded that the finite-difference scheme of Eq. (8.3.49) is unconditionally stable for all values of Δx and Δt. However, in practice, this scheme should not be used since the amplitude, $C(k)$, generally decreases with the increase of the discrete time variable.

We conclude this section with a demonstration that an exact, explicit finite-difference scheme does not exist for the free-partial Schrödinger equation.

First, we require that the discrete model satisfy the two restrictions:

(i) The discrete time-derivative be of first-order.

(ii) The discrete space-derivative be a centered, second-order expression.

Since Eq. (8.3.1) is a linear partial differential equation having only two terms, it can always be solved by the method of separation of variables. The corresponding finite-difference scheme must also have this property. Therefore, any proposed exact finite-difference scheme can be "checked" by also calculating its solutions by the method of separation of variables [27]. If the discrete model is an exact scheme, then we will find that

$$u_m^k = u(x_m, t_k) \qquad (8.3.54)$$

where u_m^k is the solution of the finite-difference equation and $u(x,t)$ is the solution to the differential equation.

The substitution of

$$u(x,t) = X(x)T(t) \qquad (8.3.55)$$

into Eq. (8.3.1) gives

$$\frac{X''}{X} = \frac{T'}{iT} = -i\omega^2, \tag{8.3.56}$$

where the primes denote the order of the indicated derivatives and ω^2 is the separation constant. For given ω^2, the special solution $u_\omega(x,t)$ is

$$u_\omega(x,t) = A(\omega)e^{i\omega(x-\omega t)}, \tag{8.3.57}$$

and the general solution is given by "summation," i.e.,

$$u(x,t) = \sum\int u_\omega(x,t)d\omega. \tag{8.3.58}$$

The general form of an explicit scheme, for Eq. (8.3.1), that satisfies the above two requirements is

$$\frac{u_m^{k+1} - \Gamma(h,\ell^2)u_m^k}{ih\Phi(h,\ell^2)} = \frac{(u_{m+1}^k + u_{m-1}^k)P(h,\ell^2) - 2Q(h,\ell^2)u_m^k}{\ell^2\Sigma(h,\ell^2)}, \tag{8.3.59}$$

where $h = \Delta t$, $\ell = \Delta x$, and the functions appearing in Eq. (8.3.59) have the representations:

$$\Phi(h,\ell^2) = 1 + h\Phi_1(h,\ell^2), \tag{8.3.60}$$

$$\Sigma(h,\ell^2) = 1 + \ell^2\Sigma_1(h,\ell^2), \tag{8.3.61}$$

$$\Gamma(h,\ell^2) = 1 + h\Gamma_1(h,\ell^2), \tag{8.3.62}$$

$$P(h,\ell^2) = 1 + \ell^2 P_1(h,\ell^2), \tag{8.3.63}$$

$$Q(h,\ell^2) = 1 + \ell^2 Q_1(h,\ell^2). \tag{8.3.64}$$

Note that $\Phi(h,\ell^2)$ may be a complex-valued function. Equation (8.3.59) can be rewritten as

$$u_m^{k+1} = A(u_{m+1}^k + u_{m-1}^k) + Bu_m^k, \tag{8.3.65}$$

where

$$A = \frac{ih\Phi P}{\ell^2\Sigma}, \tag{8.3.66}$$

$$B = \Gamma - \frac{2i\Phi Q}{\ell^2 \Sigma}. \tag{8.3.67}$$

Now assume a separation-of-variables solution

$$u_m^k = C^k D_m, \tag{8.3.68}$$

and substitute this into Eq. (8.3.65) to obtain

$$\frac{C^{k+1}}{C^k} = \frac{A(D_{m+1} + D_{m-1}) + BD_m}{D_m} = \alpha, \tag{8.3.69}$$

where α is the separation constant. Now require that C^k has the form

$$C^k = e^{i\omega^2(\Delta t)k}. \tag{8.3.70}$$

This means that the separation constant α is

$$\alpha = e^{i\omega^2 h}, \qquad h = \Delta t. \tag{8.3.71}$$

Therefore, D_m satisfies the following second-order linear difference equation

$$D_{m+1} + 2\left[\frac{B - e^{i\omega^2 h}}{2A}\right] D_m + D_{m-1} = 0. \tag{8.3.72}$$

For an exact scheme, this equation must have the solution

$$D_m = e^{i\omega(\Delta x)m}. \tag{8.3.73}$$

However, this requires that

$$\frac{B - e^{i\omega^2 h}}{2A} = -\cos(\omega\ell), \qquad \ell = \Delta x. \tag{8.3.74}$$

Written out in detail, the last equation is

$$\frac{\ell^2 \Sigma e^{i\omega^2 h} + 2ih\Phi Q - \ell^2 \Gamma \Sigma}{2ih\Phi P} = \cos(\omega\ell). \tag{8.3.75}$$

In general, for arbitrary values of ω^2, the two sides of this expression are not equal. Consequently, we conclude that the free-particle Schrödinger equation does not have an exact, explicit finite-difference scheme [28].

8.3.2 $u_t = i[u_{xx} + f(x)u]$

The time-dependent Schrödinger equation is

$$\frac{\partial u}{i \partial t} = \frac{\partial^2 u}{\partial x^2} + f(x)u. \tag{8.3.76}$$

Based on the discussion of the previous section, we expect a discrete model for Eq. (8.3.76) to take the form [8, 29]

$$\frac{u_m^{k+1} - u_m^k}{D_1(\lambda, \Delta t, \Delta x, f_m)} = \frac{u_{m+1}^k - 2u_m^k + u_{m-1}^k}{D_2(\lambda, \Delta t, \Delta x, f_m)} + f_m u_m^k, \tag{8.3.77}$$

where

$$f_m = f(x_m), \qquad x_m = (\Delta x)m, \tag{8.3.78}$$

and λ is to be selected such that the scheme is conditionally stable. The following denominator functions are suitable for this purpose:

$$D_1 = e^{i\lambda(\Delta t)} \left[\frac{e^{i(\Delta t)f_m} - 1}{f_m} \right], \tag{8.3.79}$$

$$D_2 = \left(\frac{4}{f_m} \right) \sin^2 \left[\frac{(\Delta x)\sqrt{f_m}}{2} \right]. \tag{8.3.80}$$

Examination of this scheme shows that it has the following properties:

(i) The discrete time-independent part of the scheme,

$$\frac{u_{m+1} - 2u_m + u_{m-1}}{\left(\frac{4}{f_m} \right) \sin^2 \left[\frac{(\Delta x)\sqrt{f_m}}{2} \right]} + f_m u_m = 0, \tag{8.3.81}$$

reduces to the best difference scheme for the ordinary differential equation

$$\frac{d^2 u}{dx^2} + f(x)u = 0. \tag{8.3.82}$$

(See Section 8.2.2.)

(ii) Taking the limit

$$f_m \to 0, \tag{8.3.83}$$

gives

$$\frac{u_m^{k+1} - u_m^k}{i(\Delta t)e^{i\lambda(\Delta t)}} = \frac{u_{m+1}^k - 2u_m^k + u_{m-1}^k}{(\Delta x)^2}, \tag{8.3.84}$$

which is a best difference scheme for

$$\frac{\partial u}{i\partial t} = \frac{\partial^2 u}{\partial x^2}. \tag{8.3.85}$$

The circle condition of Section 8.3.1 allows us to determine that this scheme is conditionally stable if, for example,

$$\Delta t = \frac{(\Delta x)^2}{2\sqrt{2}}, \qquad \lambda = \frac{7\pi}{\sqrt{2}(\Delta x)^2}. \tag{8.3.86}$$

Note that

$$\lambda(\Delta t) = \frac{7\pi}{4}, \tag{8.3.87}$$

and

$$e^{i\lambda(\Delta t)} = e^{-i\pi/4} = \frac{1-i}{\sqrt{2}}. \tag{8.3.88}$$

We assume that these conditions hold for the full finite-difference scheme given by Eqs. (8.3.77), (8.3.79) and (8.3.80).

Other functional forms can be chosen for the discrete-time denominator function D_1. However, the $f_m \to 0$ limit will always give relations similar to those stated in Eqs. (8.3.86) and (8.3.87).

8.3.3 Nonlinear, Cubic Schrödinger Equation

There exists a vast literature on both the properties and the numerical integration of the nonlinear, cubic Schrödinger partial differential equation [30–35]. This equation has the form

$$\frac{\partial u}{i\partial t} = \frac{\partial^2 u}{\partial x^2} + |u|^2 u, \tag{8.3.89}$$

212

and describes the asymptotic limiting behavior of a slowly varying dispersive wave envelope traveling in a nonlinear medium. Our purpose in this section is to construct a nonstandard, explicit finite-difference scheme for Eq. (8.3.89).

We begin by again considering the Duffing ordinary differential equation

$$\frac{d^2u}{dx^2} + \lambda u + u^3 = 0.\tag{8.3.90}$$

A best finite-difference scheme for it is (see Section 5.3)

$$\frac{u_{m+1} - 2u_m + u_{m-1}}{\left(\frac{4}{\lambda}\right)\sin^2\left[\frac{(\Delta x)\sqrt{\lambda}}{2}\right]} + \lambda u_m + u_m^2\left(\frac{u_{m+1} + u_{m-1}}{2}\right) = 0.\tag{8.3.91}$$

Taking the limit $\lambda \to 0$, we obtain

$$\frac{u_{m+1} - 2u_m + u_{m-1}}{(\Delta x)^2} + u_m^2\left(\frac{u_{m+1} + u_{m-1}}{2}\right) = 0,\tag{8.3.92}$$

as a best scheme for the equation

$$\frac{d^2u}{dx^2} + u^3 = 0.\tag{8.3.93}$$

These results suggest that for the differential equation

$$\frac{d^2u}{dx^2} + |u|^2u = 0,\tag{8.3.94}$$

where u is now a complex-valued function, a possible best scheme is the expression

$$\frac{u_{m+1} - 2u_m + u_{m-1}}{(\Delta x)^2} + u_m u_m^*\left(\frac{u_{m+1} + u_{m-1}}{2}\right) = 0.\tag{8.3.95}$$

Combining all these results, we obtain the following explicit finite-difference scheme for the nonlinear, cubic Schrödinger equation

$$\frac{u_m^{k+1} - u_m^k}{iD_1(\Delta t, \Delta x, \lambda)} = \frac{u_{m+1}^k - 2u_m^k + u_{m-1}^k}{(\Delta x)^2} + (u_m^k)^*\left(\frac{u_{m+1}^k + u_{m-1}^k}{2}\right)u_m^{k+1},\tag{8.3.96}$$

where D_1 is a suitable denominator function. If D_1 is selected, such that the linear part of Eq. (8.3.96) satisfies the circle condition, then from Eqs. (8.3.35) and (8.3.36)

$$iD_1 = \left(\frac{1+i}{4}\right)(\Delta x)^2, \qquad (8.3.97)$$

$$\Delta t = \frac{(\Delta x)^2}{4}, \qquad (8.3.98)$$

and Eq. (8.3.96) becomes

$$u_m^{k+1} = \frac{\beta(u_{m+1}^k + u_{m-1}^k) + (1 - 2\beta)u_m^k}{1 - \beta(\Delta x)^2(u_m^k)^*\left[\frac{u_{m+1}^k + u_{m-1}^k}{2}\right]}, \qquad (8.3.99)$$

$$\beta = \frac{1+i}{4}. \qquad (8.3.100)$$

In summary, the discrete model of the nonlinear, cubic Schrödinger equation given by Eq. (8.3.96) or Eqs. (8.3.99) and (8.3.100), is constructed by applying the nonstandard modeling rules as presented and discussed in Sections 3.4 and 3.5. These rules lead to an essentially unique structure for the finite-difference scheme.

References

1. E. Merzbacher, *Quantum Mechanics* (Wiley, New York, 1961).

2. F. Herman and E. Skillman, *Atomic Structure Calculations* (Prentice-Hall; Englewood Cliffs, NJ; 1963).

3. L. Brekhovskikh and Yu. Lupanov, *Fundamentals of Ocean Acoustics* (Springer-Verlag, New York, 1982).

4. A. K. Ghatak and K. Thyagarajan, *Contemporary Optics* (Plenum, New York, 1978).

5. P. B. Kahn, *Mathematical Methods for Scientists and Engineers: Linear and Nonlinear Systems* (Wiley-Interscience, New York, 1990).

6. C. M. Bender and S. A. Orszag, *Advanced Mathematical Methods for Scientists and Engineering* (McGraw-Hill, New York, 1978).

214

7. R. E. Mickens, *Physical Review A* **39**, 5508–5511 (1989). Stable explicit schemes for equations of Schrödinger type.

8. R. E. Mickens, Construction of stable explicit finite-difference schemes for Schrödinger type differential equations, in *Computational Acoustics* – Volume I, D. Lee, A. Cakmak and R. Vichnevetsky, editors (Elsevier, Amsterdam, 1990), pp. 11–16.

9. R. E. Mickens and I. Ramadhani, *Physical Review A* **45**, 2074–2075 (1992). Finite-difference scheme for the numerical solution of the Schrödinger equation.

10. S. E. Koonin, *Computational Physics* (Addison-Wesley; Redwood City, CA; 1986). See Section 3.1.

11. R. Chen, Z. Xu and L. Sun, *Physical Review E* **47**, 3799–3802 (1993). Finite-difference scheme to solve Schrödinger equations.

12. L. Gr. Ixaru and M. Rizea, *Journal of Computational Physics* **73**, 306–324 (1987). Numerov method maximally adapted to the Schrödinger equation.

13. A. D. Raptis and J. R. Cash, *Computer Physics Communications* **44**, 95–103 (1987). Exponential and Bessel fitting methods for the numerical solution of the Schrödinger equation.

14. F. B. Hildebrand, *Finite-Difference Equations and Simulations* (Prentice-Hall; Englewood Cliffs, NJ; 1968).

15. R. E. Mickens, Mathematical modeling of differential equations by difference equations, in *Proceedings of the First IMAC Conference on Computational Acoustics*, D. Lee, R. L. Steingberg and M. H. Schultz, editors (North-Holland, Amsterdam, 1987), pp. 387–393.

16. R. E. Mickens, *Numerical Methods for Partial Differential Equations* **2**, 313–325 (1989). Exact solutions to a finite-difference model for a nonlinear reaction-advection equation: Implications for numerical analysis.

17. R. B. Dingle and G. J. Morgan, *Applied Scientific Research* **18**, 221 (1967). WKB methods for differential equations.

18. R. E. Mickens and I. Ramadhani, WKB procedures for Schrödinger type difference equations, to appear in *Proceedings of the First World Congress of Nonlinear Analysts* (Tampa, FL; August 19–26, 1992).

19. T. F. Chan, D. Lee and L. Shen, *SIAM Journal of Numerical Analysis* **23**, 274–281 (1986). Stable explicit schemes for equations of the Schrödinger type.

20. D. Lee and S. T. McDaniel, *Ocean Acoustic Propagation by Finite Difference Methods* (Pergamon, Oxford, 1988).

21. T. F. Chan and L. Shen, *SIAM Journal of Numerical Anlaysis* **24**, 336–349 (1987). Stability analysis of difference schemes for variable coefficient Schrödinger type equations.

22. P. K. Chattaraj, S. R. Koneru and B. M. Deb, *Journal of Computational Physics* **72**, 504–512 (1987). Stability analysis of finite difference schemes for quantum mechanical equations of motion.

23. R. D. Richtmyer and K. W. Morton, *Difference Methods for Initial-Value Problems* (Interscience, New York, 1967).

24. L. Lapidus and G. F. Pinder, *Numerical Solution of Partial Differential Equations in Science and Engineering* (Wiley, New York, 1982).

25. G. D. Smith, *Numerical Solution of Partial Differential Equations: Finite Difference Methods* (Clarendon, Oxford, 2nd edition, 1978).

26. A. Goldberg, H. M. Schey and J. L. Schwartz, *American Journal of Physics* **35**, 177–186 (1967). Computer-generated motion pictures of one-dimensional quantum mechanical transmission and reflection phenomena.

27. R. E. Mickens, *Difference Equations: Theory and Applications* (Van Nostrand Reinhold, New York, 2nd edition, 1990).

28. R. E. Mickens, Proof of the impossibility of constructing exact finite-difference schemes for Schrödinger-type PDE's. Paper presented at the International Conference on Theoretical and Computational Acoustics (Mystic, Connecticut; July 5–9, 1993).

29. R. E. Mickens, *Computer Physics Communications* **63**, 203–208 (1991). Novel explicit finite-difference schemes for time-dependent Schrödinger equations.

30. J. M. Sanz-Serna and I. Christie, *Journal of Computational Physics* **67**, 348–360 (1986). A simple adaptive technique for nonlinear wave problems.

31. K. A. Ross and C. J. Thompson, *Physica* **135A**, 551–558 (1986). Iteration of some discretizations of the nonlinear Schrödinger equation.

32. B. M. Herbst and M. J. Ablowitz, *Physical Review Letters* **62**, 2065–2068 (1989). Numerically induced chaos in the nonlinear Schrödinger equation.

33. M. J. Ablowitz and H. Segur, *Solitons and the Inverse Scattering Transform* (Society for Industrial and Applied Mathematics, Philadelphia, 1981).

34. G. B. Whitham, *Linear and Nonlinear Waves* (Wiley-Interscience, New York, 1974).

35. R. K. Dodd, J. C. Eilbeck, J. D. Gibbon and H. C. Morris, *Solitons and Nonlinear Wave Equations* (Academic Press, New York, 1982).

Chapter 9
SUMMARY AND DISCUSSION

9.1 Résumé

In Chapter 1, we introduced and discussed the basic reasons for the need to construct discrete models of differential equations [1, 2]. Using standard modeling rules, finite-difference schemes were constructed for a variety of elementary, but, important ordinary and partial differential equations: the decay, Logistic, harmonic oscillator, unidirectional wave, diffusion, and Burgers' equations. A central feature of this investigation was the fundamental ambiguity in the modeling process, i.e., for a given differential equation, the use of the standard rules lead to a number of possible discrete representations and, in general, there is no *a priori* reason to select one model over another.

The genesis of numerical instabilities was presented in Chapter 2 through an examination of the solution behaviors of finite-difference models of the above mentioned differential equations. Numerical instabilities are solutions of the finite-difference equations that do not correspond to any solutions of the original differential equation. Our study showed that numerical instabilities could occur under a variety of circumstances for discrete models constructed according to the standard rules.

A partial resolution to the problem of numerical instabilities was given in Chapter 3. This chapter introduced the notion of an *exact* finite-difference scheme. It was shown, by means of a theorem, that, in general, ordinary differential equations have exact finite-difference equation representations. This theorem was then used to construct exact discrete models for several differential equations. A study of these exact schemes then led to the formulation of a set of nonstandard modeling rules [3]. For a particular differential equation, the construction of a finite-difference scheme based on these rules gives what is called a *best* finite-difference model [3].

218

The remainder of the book applied the nonstandard rules to an assortment of special classes of both ordinary and partial differential equations. For example, Chapter 4 details the construction of discrete representations for a single scalar ordinary differential equation

$$\frac{dy}{dt} = f(y), \tag{9.1.1}$$

such that the linear stability properties of the fixed-points of the finite-difference scheme are exactly the same as the corresponding fixed-points of the differential equation for all values of the step-size, $h = \Delta t$. This result eliminates all the elementary numerical instabilities and is based on the idea of using a renormalized "denominator function" [3, 4]. Likewise, in Chapter 5, a detailed study is made on the construction of best finite-difference schemes for nonlinear, second-order oscillator differential equations. The equations examined included conservative, damped, and limit-cycle oscillators.

Chapter 6 investigated discrete models for two first-order, coupled ordinary differential equations. The equations considered have only a single fixed-point. Our best discrete models were constructed such that the linear stability properties of the finite-difference equations were exactly the same as the corresponding differential equations. A major discovery was the realization that only the semi-explicit scheme gave results consistent with other best discrete representations. (A semi-explicit scheme is an explicit discrete model for which the dependent variables have to be calculated in a definite order.)

In Chapter 7, we constructed best finite-difference schemes for linear and nonlinear wave, diffusion, and Burgers' type partial differential equations. In general, the time derivatives were of first-order and the nonlinearities were quadratic in the dependent variable and its space derivatives. The basic technique used in these constructions was to consider the various sub-equations of a given partial differential equation, construct the exact or best discrete models for the sub-equations, and then combine all the sub-equations to obtain a discrete model for the full partial differential equation. We found that for certain equations both explicit and implicit discrete representations could be found. However, other equations lead to only implicit schemes. Another important finding was that for those partial differential equations for which exact schemes could be constructed, there was always a definite functional relation between the time and space step-sizes.

Discrete models of Schrödinger ordinary and partial differential equations were studied in Chapter 8. One major result discussed in this chapter was the combining of the standard Numerov technique with the Mickens-Ramadhani scheme to obtain a new and improved finite-difference scheme for Schrödinger type ordinary differential equations [5]

$$\frac{d^2y}{dx^2} + f(x)y = 0. \tag{9.1.2}$$

Another significant result was the use of nonstandard denominator functions to construct explicit, forward Euler type schemes for the time-dependent Schrödinger partial differential equation [6]

$$\frac{\partial u}{i\partial t} = \frac{\partial^2 u}{\partial x^2} + f(x)u. \tag{9.1.3}$$

In the next section, we will restate the nonstandard modeling rules given in Section 3.4 and discuss the difficulties of applying them in the actual construction of best schemes. These difficulties and successes will be illustrated by way of two examples in Section 9.3. Finally, in Section 9.4, we present a number of unresolved issues and further directions for research.

9.2 Nonstandard Modeling Rules Revisited

We now restate and discuss the five nonstandard rules for constructing finite-difference models of differential equations as given in Section 3.4.

Rule 1. The orders of the discrete derivatives must be exactly equal to the orders of the corresponding derivatives of the differential equation.

Discussion 1. Numerical and analytical experience with discrete models that violate this rule show that, in general, when the orders of the discrete derivatives are larger than the corresponding orders that appear in the differential equation, numerical instabilities in the form of oscillations often appear. Depending on the particular differential equation, these oscillations may be bounded or unbounded. The mathematical reason for their occurrence is that the discrete equations have a larger class of solutions than the differential equation. For example, the linear differential equation

$$\frac{dy}{dt} = -y, \tag{9.2.1}$$

if modeled by a central difference scheme

$$\frac{y_{k+1} - y_{k-1}}{2h} = -y_k \tag{9.2.2}$$

has an extra "ghost" solution [7, 8, 9] because Eq. (9.2.2) is of second-order and, consequently, has two linearly independent solutions, while Eq. (9.2.1) has only one solution. Thus, violation of this rule automatically assures the existence of numerical instabilities. In general, this rule applies to each individual derivative in the various terms of a differential equation.

Rule 2. Denominator functions for the discrete derivatives must, in general, be expressed in terms of more complicated functions of the step-sizes than those conventionally used.

Discussion 2. It is not *a priori* obvious that the denominator function for the Logistic differential equation

$$\frac{dy}{dt} = y(1 - y), \tag{9.2.3}$$

is

$$D_1 = e^h - 1, \qquad h = \Delta t, \tag{9.2.4}$$

or that the denominator function for the harmonic oscillation equation is

$$D_2 = 4\sin^2\left(\frac{h}{2}\right), \qquad h = \Delta t, \tag{9.2.5}$$

where the corresponding exact finite-difference schemes, resepectively, are

$$\frac{y_{k+1} - y_k}{D_1(h)} = y_k(1 - y_{k+1}), \tag{9.2.6}$$

$$\frac{y_{k+1} - 2y_k + y_{k-1}}{D_2(h)} + y_k = 0. \tag{9.2.7}$$

At the present stage in the development of best schemes, the selection of an appropriate denominator function is an "art." However, close examination of differential equations for which exact schemes are known, shows that the denominator functions generally are functions that are related to particular solutions or properties of the general solution to the differential equation. For example, the Logistic equation has exponential behavior for its solutions that start near the fixed-point at $y(t) = 0$. Likewise, the harmonic oscillator equation has the solution $\sin(t)$. These functions indicate the general difficulty of selecting a denominator function for a discrete derivative for the case where we have no knowledge of the behavior of the solutions to the differential equation. Therefore, this result places great importance on the necessity of the modeler to obtain as much analytic knowledge as possible about the solutions to the differential equation.

Rule 3. Nonlinear terms must, in general, be modeled nonlocally on the computational grid or lattice.

Discussion 3. The particular way that this occurs will depend, of course, on the exact nature of the nonlinear term and the order(s) of the differential equation under consideration. For example, terms of the form y^2 and y^3 which appear in a first-order, nonlinear ordinary differential equation could be modeled in the following way

$$y^2 \rightarrow y_{k+1}y_k, \tag{9.2.8}$$

$$y^3 \rightarrow y_{k+1} y_k \left(\frac{y_{k+1} + y_k}{2} \right). \tag{9.2.9}$$

Note that the y^3 term could also be presented by one of the forms

$$y^3 \rightarrow \begin{cases} y_k^2 y_{k+1}, \\ y_k^2 \left(\frac{y_{k+1} + y_k}{2} \right). \end{cases} \tag{9.2.10}$$

The particular form selected, from Eqs. (9.2.9) and (9.2.10), will depend on what other conditions need to be satisfied for the full discrete model.

Rule 4. Special solutions of the differential equation should also be special (discrete) solutions of the finite-difference models.

Discussion 4. The violation of this rule indicates that the discrete model is not a good representation of the differential equation. For simple special solutions, like fixed-points or constant solutions, this rule is generally automatically satisfied. However, it is for the more complicated special solutions, like rational functions, that this rule is violated. For example, the Burgers' partial differential equation

$$u_t + u u_x = 0, \tag{9.2.11}$$

has a special rational solution

$$u(x,t) = \frac{\alpha x + A_2}{\alpha t + A_1}, \tag{9.2.12}$$

where (α, A_1, A_2) are arbitrary constants. However, the discrete model

$$\frac{u_m^{k+1} - u_m^k}{\Delta t} + u_m^k \left(\frac{u_{m+1}^k - u_m^k}{\Delta x} \right) = 0, \tag{9.2.13}$$

does not have this rational solution, while the following finite-difference model does:

$$\frac{u_m^{k+1} - u_m^k}{\Delta t} + u_m^{k+1} \left(\frac{u_m^k - u_{m-1}^k}{\Delta x} \right) = 0. \tag{9.2.14}$$

Rule 5. The finite-difference equations should not have solutions that do not correspond to solutions of the differential equations.

Discussion 5. This rule is not quite the "inverse" of Rule 4. The best way to illustrate this rule is by way of an example. Consider the discrete model given by Eq. (9.2.13). If we seek a special solution having the form [10]

$$u_m^k = C^k D_m, \qquad (9.2.15)$$

then C^k and D_m satisfy the equations

$$C^{k+1} = C^k - \alpha(\Delta t)(C^k)^2, \qquad (9.2.16)$$

$$D_{m+1} = D_m + \alpha(\Delta x), \qquad (9.2.17)$$

where α is an arbitrary separation constant. The second equation can be solved to give for D_m the result

$$D_m = \alpha(\Delta x)m + A_2 = \alpha x_m + A_2, \qquad (9.2.18)$$

where A_2 is an arbitrary constant. This is just the discrete form of the numerator on the right-side of Eq. (9.2.12). However, Eq. (9.2.16) is the Logistic difference equation and has no solution corresponding to the discrete version of the denominator on the right-side of Eq. (9.2.12) [10]. Thus, we conclude that the finite-difference scheme given by Eq. (9.2.13) cannot be a best scheme. In many situations, the consequences of Rule 5 are related to the violation of one of the previous four rules. In the case just considered, Rule 3 is not satisfied.

In the next section, we again illustrate the use of the nonstandard modeling rules by applying them to two non-trivial differential equations.

9.3 Two Examples

The two investigations to follow on the construction of discrete models for particular differential equations will clearly show the advantages and pitfalls in the procedures for constructing best finite-difference schemes with our present knowledge of the nonstandard modeling rules.

The first equation to be considered is the ordinary differential equation satisfied by the Weierstrass elliptic function $\mathcal{P}(z)$ [11]. In first- and second-order forms, these equations are

$$\left(\frac{dp}{dz}\right)^2 = 4p^3 - g_2 p - g_3, \tag{9.3.1}$$

$$\frac{d^2 p}{dz^2} = 6p^2 - \left(\frac{1}{2}\right) g_2, \tag{9.3.2}$$

where the constants g_2 and g_3 are called the "invariants." Question: With just this information, what are best finite-difference schemes for these equations?

To begin, we construct a best scheme for Eq. (9.3.1) and then obtain the best scheme for its second-order form by differencing this expression. Based on the results of Sections 5.3 and 5.5, we obtain for the discrete version of Eq. (9.3.1) the result

$$\frac{(p_{m+1} - p_m)^2}{\psi(h)} = 4p_{m+1} p_m \left(\frac{p_{m+1} + p_m}{2}\right) - g_2 \left(\frac{p_{m+1} + p_m}{2}\right) - g_3, \tag{9.3.3}$$

where $\psi(h)$ is any function with the property

$$\psi(h) = h^2 + O(h^4), \qquad h = \Delta z, \tag{9.3.4}$$

and

$$z_m = (\Delta z)m, \qquad p_m = p(z_m). \tag{9.3.5}$$

Differencing Eq. (9.3.3) and eliminating common factors gives

$$\frac{p_{m+1} - 2p_m + p_{m-1}}{\psi(h)} = 6p_m \left(\frac{p_{m+1} + p_m + p_{m-1}}{3}\right) - \left(\frac{1}{2}\right) g_2. \tag{9.3.6}$$

Without additional information, we must stop at this stage of the modeling process. Therefore, the use of the nonstandard modeling rules give Eqs. (9.3.3) and (9.3.6) as the best finite-difference schemes for, respectively, Eqs. (9.3.1) and (9.3.2).

The application of the standard modeling rules would give the following discrete expressions for Eqs. (9.3.1) and (9.3.2):

$$\frac{(p_{m+1} - p_m)^2}{h^2} = 4p_m^3 - g_2 p_m - g_3, \tag{9.3.7}$$

$$\frac{p_{m+1} - 2p_m + p_{m-1}}{h^2} = 6p_m^2 - \left(\frac{1}{2}\right)g_2. \tag{9.3.8}$$

Note that differencing Eq. (9.3.7) does not give Eq. (9.3.8). What is obtained is the expression

$$\frac{p_{m+1} - 2p_m + p_{m-1}}{h^2} = \left[6\left(\frac{p_m^2 + p_m p_{m-1} + p_{m-1}^2}{3}\right) - \left(\frac{1}{2}\right)g_2\right]$$
$$\cdot \left[\frac{2(p_m - p_{m-1})}{(p_{m+1} - p_{m-1})}\right]. \tag{9.3.9}$$

Of great value to us is the fact that Potts [12] has obtained the exact finite-difference schemes for Eqs. (9.3.1) and (9.3.2). They are

$$\frac{(p_{m+1} - p_m)^2}{\phi(h)} = 4p_{m+1}p_m\left(\frac{p_{m+1} + p_m}{2}\right)$$
$$- g_2\left(\frac{p_{m+1} + p_m}{2}\right) - g_3$$
$$- \phi(h)\left\{\left[p_{m+1}p_m + \left(\frac{1}{2}\right)g_2\right]^2 + g_3(p_{m+1} + p_m)\right\} \tag{9.3.10}$$

and

$$\frac{p_{m+1} - 2p_m + p_{m-1}}{\phi(h)} = 6p_m\left(\frac{p_{m+1} + p_m + p_{m-1}}{3}\right) - \left(\frac{1}{2}\right)g_2$$
$$- \phi(h)\left\{p_m^2(p_{m+1} + p_{m-1}) + \left(\frac{1}{2}\right)g_2 p_m + g_3\right\}, \tag{9.3.11}$$

where the denominator function $\phi(h)$ is

$$\phi(h) = \frac{1}{\mathcal{P}(h)} = h^2 + O(h^4), \tag{9.3.12}$$

and $\mathcal{P}(z)$ is the Weierstrass elliptic function. The right-sides of these equations, except for the last bracketed term, agree with the results obtained using the non-standard modeling rules. It is difficult to see how the bracketed terms could be discovered with just a knowledge of the five rules. This illustrates a general problem that occurs in the discrete modeling of differential equations: It is always possible to add terms to a finite-difference scheme that are proportional to high powers of the

226

step-size. These new expressions converge to the same differential equation when the step-size is taken to zero.

The exact denominator function, Eq. (9.3.12), is equal to the inverse of the Weierstrass elliptic function, $\mathcal{P}(z)$, evaluated at $z = h$. Again, no current best finite-difference model would give this result which clearly depends on knowing the solution to the original nonlinear, second-order differential equation.

The second equation to be considered is the modified Korteweg-de-Vries partial differential equation which is usually written as [13, 14]

$$\phi_t + 6\phi^2\phi_x + \phi_{xxx} = 0. \tag{9.3.13}$$

The t-independent part of this equation

$$\phi_{xxx} + 6\phi^2\phi_x = 0, \tag{9.3.14}$$

can be integrated once to give

$$\phi_{xx} + 2\phi^3 = A, \tag{9.3.15}$$

and integrated a second time to give

$$\left(\frac{1}{2}\right)(\phi_x)^2 + \left(\frac{1}{2}\right)\phi^4 = A\phi + B, \tag{9.3.16}$$

where A and B are arbitrary integration constants. As in the previous example, we will construct a best finite-difference scheme for Eq. (9.3.16) and then apply the difference operator to it twice to obtain the best scheme for Eq. (9.3.14). Thus, for Eq. (9.3.16), we have

$$\left(\frac{1}{2}\right)\left[\frac{\phi_m - \phi_{m-1}}{D_1(h)}\right]^2 + \left(\frac{1}{2}\right)\phi_m^2\phi_{m-1}^2 = A\left(\frac{\phi_m + \phi_{m-1}}{2}\right) + B, \tag{9.3.17}$$

where

$$D_1(h) = h + O(h^2), \tag{9.3.18}$$

$$x_m = (\Delta x)m, \qquad \Delta x = h. \tag{9.3.19}$$

Differencing Eq. (9.3.17) gives for the discrete form of Eq. (9.3.15) the expression

$$\frac{\phi_{m+1} - 2\phi_m + \phi_{m-1}}{[D_1(h)]^2} + 2\phi_m^2\left(\frac{\phi_{m+1} + \phi_{m-1}}{2}\right) = A. \tag{9.3.20}$$

Finally, differencing Eq. (9.3.20) gives

$$\frac{\phi_{m+2} - 3\phi_{m+1} + 3\phi_m - \phi_{m-1}}{[D_1(h)]^2 D_2(h)}$$
$$+ 6\left[\frac{\phi_{m+1}^2(\phi_{m+2} + \phi_m) - \phi_m^2(\phi_{m+1} + \phi_{m-1})}{6D_2(h)}\right] = 0, \tag{9.3.21}$$

where

$$D_2(h) = h + O(h^2). \tag{9.3.22}$$

Note that the denominator functions $D_1(h)$ and $D_2(h)$ need not be equal to each other. At this level of the analysis, it is only required that they be functions that satisfy the conditions of Eqs. (9.3.18) and (9.3.22). Clearly, the best finite-difference scheme for Eq. (9.3.14), as given by Eq. (9.3.21), is not the one that would come from the use of the standard modeling rules. Also, observe that ϕ_{m+2} can be expressed in terms of $(\phi_{m+1}, \phi_m, \phi_{m-1})$ since Eq. (9.3.21) is linear in ϕ_{m+2}.

We can now integrate the result of Eq. (9.3.21) into a full discrete model for the Korteweg-de-Vries equation. If we require that the nonlinear terms of the scheme be nonlocal in the discrete time levels, then the simplest finite-difference scheme is

$$\frac{\phi_m^{k+1} - \phi_m^k}{D_3(\Delta t)} + 6\left[\frac{(\phi_{m+1}^k)^2(\phi_{m+2}^{k+1} - \phi_m^{k+1}) - (\phi_m^k)^2(\phi_{m+1}^{k+1} + \phi_{m-1}^{k+1})}{6D_2(\Delta x)}\right]$$
$$+ \frac{\phi_{m+2}^{k+1} - 3\phi_{m+1}^{k+1} + 3\phi_m^{k+1} - \phi_{m-1}^{k+1}}{[D_1(\Delta x)]^2 D_2(\Delta x)} = 0, \tag{9.3.23}$$

where

$$D_3(\Delta t) = \Delta t + O[(\Delta t)^2]. \tag{9.3.24}$$

Observe that this scheme is implicit; however, it is linear in all the terms that are evaluated on the $(k+1)$-th discrete-time level.

Other nonstandard finite-difference models for the Korteweg-de-Vries equation can be established by applying the modeling rules directly to the partial differential equation written in the form

$$\phi_t + 3\phi(\phi^2)_x + \phi_{xxx} = 0. \tag{9.3.25}$$

The result is the following expression

$$\frac{\phi_m^{k+1} - \phi_m^k}{D_3(\Delta t)} + 3\phi_m^{k+1}\left[\frac{(\phi_m^k)^2 - (\phi_{m-1}^k)^2}{D_2(\Delta x)}\right]$$
$$+ \frac{\phi_{m+2}^k - 3\phi_{m+1}^k + 3\phi_m^k - \phi_{m-1}^k}{[D_1(\Delta x)]^2 D_2(\Delta x)} = 0, \tag{9.3.26}$$

which is an explicit scheme. The corresponding implicit scheme is

$$\frac{\phi_m^{k+1} - \phi_m^k}{D_3(\Delta t)} + 3\phi_m^{k+1}\left[\frac{(\phi_m^k)^2 - (\phi_{m-1}^k)^2}{D_2(\Delta x)}\right]$$
$$+ \frac{\phi_{m+2}^{k+1} - 3\phi_{m+1}^{k+1} + 3\phi_m^{k+1} - \phi_{m-1}^{k+1}}{[D_1(\Delta x)]^2 D_2(\Delta x)} = 0. \tag{9.3.27}$$

For actual numerical work, in the absence of additional information, the following choices can be made for the denominator functions

$$D_1(\Delta x) = \Delta x, \qquad D_2(\Delta x) = \Delta x, \qquad D_3(\Delta t) = \Delta t. \tag{9.3.28}$$

Note that the above construction processes do not give a functional relation between the space and time step-sizes. If such a relation exists, it will have to be determined from other considerations, the nature of which is not known at the present time.

It is mildly disturbing to see two such radically different types of discrete models appear for the Korteweg-de-Vries equation, i.e., Eq. (9.3.23) and either Eq. (9.3.26) or Eq. (9.3.27). Both types of models seemingly come from the direct application of the nonstandard modeling rules. For Eq. (9.3.23), we first obtained a best finite-difference scheme for the t-independent part of the partial differential equation and then required that the full discrete model contain it in an appropriate fashion. The other models, Eqs. (9.3.26) and (9.3.27), came directly from the application of the nonstandard rules to a modified version for the partial differential equation. From a fundamental point of view, Eqs. (9.3.26) and (9.3.27) satisfy the requirement that the nonlinear term, $3\phi(\phi^2)_x$, is modeled by a form that contains a first-order discrete derivative. The corresponding term for Eq. (9.3.23) does not have this property.

9.4 Future Directions

This book has given a summary of the author's work to date on the formulation and application of nonstandard rules for constructing discrete models of differential equations. The on-going research in this area involves a detailed study of several critical issues. In particular, the following questions/topics are being actively pursued:

(1) Can the modeling rules be generalized to partial differentials in a higher number of space-dimensions?

(2) Do additional nonstandard modeling rules exist? Must the modeling rules always be satisfied in their application to the construction of best finite-difference schemes? If there are exceptions, then when do they occur?

(3) For a given differential equation, in the absence of knowledge about the exact solution, how should the denominator functions that appear in the discrete derivatives be selected?

(4) For discrete models of partial differential equations, do general principles exist for determining the (expected) functional relation among the various step-sizes? If such relations cannot be found, then how should the step-sizes be selected in an actual numerical calculation?

(5) Do nonlinear stability methods exist that would help us in the construction of best discrete models? Can analytical techniques be combined with finite-difference procedures to obtain "better" nonstandard discrete schemes?

(6) Can an exact finite-difference scheme be constructed for the two-dimensional Laplace equation? (This problem might have some relationship with the

area of discrete analytic functions. This topic is discussed in the papers of Duffin [15], Duffin and Duris [16, 17], Deeter and Lord [18], and Hayabara [19].)

(7) Is it possible to construct better explicit finite-difference schemes for both linear and nonlinear time-dependent Schrödinger type partial differential equations? If such is the case, then what are the conditional stability requirements?

The useful resolution of these questions will be of great interest to those persons whose scientific work is based on the numerical integration of differential equations.

References

1. M. B. Allen III, I. Herrera and G. F. Pinder, *Numerical Modeling in Science and Engineering* (Wiley-Interscience, New York, 1988).

2. D. Potter, *Computational Physics* (Wiley-Interscience, Chichester, 1973).

3. R. E. Mickens, *Numerical Methods for Partial Differential Equations* 5, 313–325 (1989). Exact solution to a finite-difference model of a nonlinear reaction-advection.

4. R. E. Mickens and A. Smith, *Journal of the Franklin Institute* 327, 143–149 (1990). Finite-difference models of ordinary differential equations: Influence of denominator functions.

5. R. Chen, Z. Xu and L. Sun, *Physical Review* 47E, 3799–3802 (1992). Finite-difference scheme to solve Schrödinger equations.

6. R. E. Mickens, A new finite-difference scheme for Schrödinger type partial differential equations, in *Computational Acoustics*, Volume 2, editors, D. Lee, R. Vichnevetsky and A. R. Robinson (North-Holland, Amsterdam, 1993), pp. 233–239.

7. F. B. Hildebrand, *Finite-Difference Equations and Simulations* (Prentice-Hall; Englewood Cliffs, NJ; 1968).

8. M. Yamajuti and H. Matano, *Proceedings of the Japan Academy* 55, 78–80 (1979). Euler's finite difference scheme and chaos.

9. S. Ushiki, *Physica* 4D, 407–424 (1982). Central difference scheme and chaos.

10. R. E. Mickens, *Difference Equations: Theory and Applications* (Van Nostrand Reinhold, New York, 1990).

11. A. Abramowitz and I. A. Stegun, editors, *Handbook of Mathematical Functions* (U.S. National Bureau of Standards; Washington, DC; 1964).

12. R. B. Potts, *Bulletin of the Australian Mathematical Society* **35**, 43–48 (1987). Weierstrass elliptic difference equations.

13. G. Eilenberger, *Solitons: Mathematical Methods for Physicists* (Springer-Verlag, Berlin, 1983).

14. G. B. Whitham, *Linear and Nonlinear Waves* (Wiley, New York, 1974).

15. R. J. Duffin, *Duke Mathematics Journal* **23**, 335–363 (1956). Basic properties of discrete analytic functions.

16. R. J. Duffin and C. S. Duris, *Duke Mathematics Journal* **31**, 199–220 (1964). Convolution products for discrete function theory.

17. R. J. Duffin and C. S. Duris, *Journal of Mathematical Analysis and Applications* **9**, 252–267 (1964). Discrete analytic continuation of solutions of difference equations.

18. C. R. Deeter and M. E. Lord, *Journal of Mathematical Analysis and Applications* **26**, 92–113 (1969). Further theory of operational calculus on discrete analytic functions.

19. S. Hayabara, *Journal of Mathematical Analysis and Applications* **34**, 339–359 (1971). Discrete analytic function theory of n-variables.

Appendix A
DIFFERENCE EQUATIONS

A.1 Linear Equations

The n-th order linear inhomogeneous difference equation with constant coefficients has the form

$$y_{k+n} + a_1 y_{k+n-1} + a_2 y_{k+n-2} + \cdots + a_n y_k = R_k, \qquad (A.1.1)$$

where the $\{a_i\}$ are given constants, with $a_n \neq 0$, and R_k is a known function of k. If $R_k = 0$, then Eq. (A.1.1) is an n-th order homogeneous difference equation, i.e.,

$$y_{k+n} + a_1 y_{k+n-1} + a_2 y_{k+n-2} + \cdots + a_n y_k = 0. \qquad (A.1.2)$$

The general solution to the homogeneous equation consists of a linear combination of n linearly independent functions $\{y_k^{(i)}\}$:

$$y_k^{(H)} = c_1 y_k^{(1)} + c_2 y_k^{(2)} + \cdots + c_n y_k^{(n)}, \qquad (A.1.3)$$

where the $\{c_i\}$ are n arbitrary constants. The n linearly independent functions $\{y_k^{(i)}\}$ are determined as follows [1]:

(i) First, construct the *characteristic equation* associated with Eq. (A.1.2); it is given by the expression

$$r^n + a_1 r^{n-1} + a_2 r^{n-2} + \cdots + a_n = 0. \qquad (A.1.4)$$

(ii) Denote the n roots of the characteristic equation by $\{r_i\}$.

(iii) The n linearly independent functions $\{y_k^{(i)}\}$ are

$$y_k^{(i)} = (r_i)^k \qquad i = (1, 2, \ldots, n). \qquad (A.1.5)$$

Consequently, the general solution to Eq. (A.1.2) is

$$y_k^{(H)} = c_1(r_1)^k + c_2(r_2)^k + \cdots + c_n(r_n)^k. \tag{A.1.6}$$

This result assumes that all the roots of the characteristic equation are distinct.

(iv) If root r_i occurs with a multiplicity $m \leq n$, then its contribution to the homogeneous solution takes the form

$$y_k^{(i)} = (A_1 + A_2 k + \cdots + A_m k^{m-1})(r_i)^k. \tag{A.1.7}$$

If $R_k \neq 0$, then the solution to the inhomogeneous Eq. (A.1.1) is a sum of the homogeneous solution $y_k^{(H)}$ and a particular solution to Eq. (A.1.1), i.e.,

$$y_k = y_k^{(H)} + y_k^{(P)}. \tag{A.1.8}$$

If R_k is a linear combination of various products of the terms

$$a^k, \quad e^{bk}, \quad \sin(ck), \quad \cos(ck), \quad k^\ell, \tag{A.1.9}$$

where (a, b, c) are constants and ℓ is a non-negative integer, then rules exist for constructing particular solutions [1]. When solutions to linear inhomogeneous equations are cited in this book, the particular solutions can usually be determined by inspection.

A.2 Riccati Equations

The following nonlinear, first-order difference equation

$$P y_{k+1} y_k + Q y_{k+1} + R y_k = S, \tag{A.2.1}$$

where (P, Q, R, S) are constants, appears in a number of places in the deliberations of this book. This equation is a special case of Riccati difference equation [1]. This equation can be solved exactly by first dividing by P and shifting the index k to give

$$y_k y_{k-1} + A y_k + B y_{k-1} = C, \qquad (A.2.2)$$

where

$$A = \frac{Q}{P}, \qquad B = \frac{R}{P}, \qquad C = \frac{S}{P}. \qquad (A.2.3)$$

The nonlinear transformation

$$y_k = \frac{x_k - B x_{k+1}}{x_{k+1}}, \qquad (A.2.4)$$

reduces Eq. (A.2.2) to the following linear, second-order equation for x_k:

$$(AB + C)x_{k+1} - (A - B)x_k - x_{k-1} = 0. \qquad (A.2.5)$$

This equation can be solved by the method of Section A.1 and consequently, the general solution to Eq. (A.2.1) can be found.

Note that if $S = 0$, Eq. (A.2.1) becomes

$$P y_{k+1} y_k + Q y_{k+1} + R y_k = 0 \qquad (A.2.6)$$

and the substitution

$$y_k = \frac{1}{x_k}, \qquad (A.2.7)$$

gives the first-order, linear difference equation

$$R x_{k+1} + Q x_k + P = 0, \qquad (A.2.8)$$

which can be easily solved.

A.3 Separation-of-Variables

Many partial difference equations with constant coefficients can be solved to obtain special solutions by the method of separation-of-variables [1].

Let $z(k,\ell)$ denote a function of the discrete (integer) variables (k,ℓ). Now define the two shift operators, E_1 and E_2, as follows

$$(E_1)^m z(k,\ell) = z(k+m,\ell),\qquad\qquad\text{(A.3.1)}$$

$$(E_2)^m z(k,\ell) = z(k,\ell+m),\qquad\qquad\text{(A.3.2)}$$

where m is an integer. Now consider the linear partial difference equation

$$\psi(E_1, E_2, k, \ell)z(k,\ell) = 0,\qquad\qquad\text{(A.3.3)}$$

where the operator ψ is a polynomial function of E_1 and E_2. The basic principle of the method of separation-of-variables is to assume that $z(k,\ell)$ can be written as

$$z(k,\ell) = C(k)D(\ell).\qquad\qquad\text{(A.3.4)}$$

Assume further that when this form is substituted into Eq. (A.3.3), an equation having the structure

$$\frac{f_1(E_1,k)C(k)}{f_2(E_1,k)C(k)} = \frac{g_1(E_2,\ell)D(\ell)}{g_2(E_2,\ell)D(\ell)},\qquad\qquad\text{(A.3.5)}$$

is obtained. Under these conditions, $C(k)$ and $D(\ell)$ satisfy the ordinary difference equations

$$f_1(E_1, k)C(k) = \alpha f_2(E_1, k)C(k),\qquad\qquad\text{(A.3.6)}$$

$$g_1(E_2, \ell)D(\ell) = \alpha g_2(E_2, \ell)D(\ell),\qquad\qquad\text{(A.3.7)}$$

where α is the arbitrary separation constant. The solutions to these equations will depend on α, i.e., $C(k,\alpha)$ and $D(\ell,\alpha)$. Therefore, the special solution $z(k,\ell)$ given by Eq. (A.3.4) will also depend on α. Since the original partial difference equation is linear, a general solution can be obtained by "summing" over α, i.e.,

$$z(k,\ell) = \sum\int z(k,\ell,\alpha)d\alpha,\qquad\qquad\text{(A.3.8)}$$

where

$$z(k,\ell,\alpha) = C(k,\alpha)D(\ell,\alpha).\qquad\qquad\text{(A.3.9)}$$

Reference

1. R. E. Mickens, *Difference Equations: Theory and Applications* (Van Nostrand Reinhold, New York, 1990). See Sections 4.2, 4.5, 5.3 and 6.3.

Appendix B
LINEAR STABILITY ANALYSIS

B.1 Ordinary Differential Equations

Consider the scalar ordinary differential equation

$$\frac{dy}{dt} = f(y). \tag{B.1.1}$$

Assume that

$$f(\bar{y}) = 0, \tag{B.1.2}$$

has m simple zeros, where m may be unbounded. The solutions of Eq. (B.1.2) are fixed-points of the differential equation and correspond to constant solutions.

The linear stability properties of the fixed-points are determined by investigating the behavior of small perturbations about a given fixed-point [1, 2]. Consider the i-th fixed-point, $\bar{y}^{(i)}$. The perturbed trajectory takes the form

$$y(t) = \bar{y}^{(i)} + \epsilon(t), \tag{B.1.3}$$

where

$$|\epsilon(t)| \ll |\bar{y}^{(i)}|. \tag{B.1.4}$$

Substitution of Eq. (B.1.3) into Eq. (B.1.1) gives

$$\frac{d\epsilon}{dt} = f[\bar{y}^{(i)}] + R_i\epsilon + O(\epsilon^2), \tag{B.1.5}$$

where

$$R_i = \frac{df}{dy}\bigg|_{y=\bar{y}^{(i)}}. \tag{B.1.6}$$

The linear stability equation is given by the linear terms of Eq. (B.1.5), i.e.,

$$\frac{d\epsilon}{dt} = R_i\epsilon. \tag{B.1.7}$$

237

The solution of this equation is

$$\epsilon(t) = \epsilon_0 e^{R_i t}. \tag{B.1.8}$$

The fixed-point, $y(t) = \bar{y}^{(i)}$, is said to be linearly stable if $R_i < 0$, and linearly unstable if $R_i > 0$.

These results can be easily generalized to the case of coupled first-order differential equations [1, 2].

B.2 Ordinary Difference Equations

The fixed-points of the first-order difference equation

$$y_{k+1} = F(y_k), \tag{B.2.1}$$

are the solutions to

$$\bar{y} = F(\bar{y}). \tag{B.2.2}$$

Assume that all the zeros of

$$G(\bar{y}) = F(\bar{y}) - \bar{y} = 0, \tag{B.2.3}$$

are simple and denote the n fixed-points by $\{\bar{y}^{(j)}\}$, $j = (1, 2, \ldots, n)$. The perturbed solution about the particular fixed-point

$$y_k = \bar{y}^{(j)} \tag{B.2.4}$$

can be written as

$$y_k = \bar{y}^{(j)} + \epsilon_k, \tag{B.2.5}$$

where

$$|\epsilon_k| \ll |\bar{y}^{(j)}|. \tag{B.2.6}$$

Substitution of Eq. (B.2.5) into Eq. (B.2.1) and retaining only linear terms in ϵ_k gives

$$\epsilon_{k+1} = R_j \epsilon_k \tag{B.2.7}$$

238

where

$$R_j = \left.\frac{dF}{dy}\right|_{y=y^{(j)}} \qquad (B.2.8)$$

Equation (B.2.7) has the solution

$$e_k = e_0(R_j)^k. \qquad (B.2.9)$$

Thus, the j-th fixed-point of Eq. (B.2.1) is said to be linearly stable if $|R_j| < 1$ and linearly unstable for $|R_j| > 0$.

Additional details and the generalization to higher-order systems of difference equations are discussed in references [3, 4, 5, 6].

References

1. L. Cesari, *Asymptotic Behavior and Stability Problems in Ordinary Differential Equations* (Academic Press, New York, 2nd edition, 1963).

2. D. A. Sánchez, *Ordinary Differential Equations and Stability Theory* (W. H. Freeman, San Francisco, 1968).

3. R. E. Mickens, *Difference Equations: Theory and Applications* (Van Nostrand Reinhold, New York, 1990). See Section 7.4.

4. J. T. Sandefur, *Discrete Dynamical Systems: Theory and Applications* (Clarendon Press, Oxford, 1990). See Chapter 4.

5. E. I. Jury, *IEEE Transactions on Automatic Control* **AC-16**, 233–240 (1971). The inners approach to some problems of system theory.

6. E. R. Lewis, *Network Models in Population Biology* (Springer-Verlag, New York, 1977).

Appendix C

DISCRETE WKB METHOD

In Section 8.2, use was made of a discrete version of the WKB method [1] to calculate the asymptotic behavior of difference equations having the form

$$y_{k+1} + y_{k-1} = 2\sigma_k y_k, \tag{C.1}$$

where σ_k has the asymptotic representation

$$\sigma_k = A_0 + \frac{A_1}{k} + \frac{A_2}{k^2} + \frac{A_3}{k^3} + O\left(\frac{1}{k^4}\right), \tag{C.2}$$

with

$$|A_0| \neq 1. \tag{C.3}$$

The asymptotic behavior of y_k is given by the expression [2]

$$y_k = k^\theta e^{B_0 k} \left[1 + \frac{B_1}{k} + \frac{B_2}{k^2} + \frac{B_3}{k^3} + O\left(\frac{1}{k^4}\right)\right], \tag{C.4}$$

where $(\theta, B_0, B_1, B_2, B_3)$ may be complex-valued and are to be determined as functions of (A_0, A_1, A_2, A_3).

If the form of Eq. (C.4) is substituted into Eq. (C.1) and use is made of the relation

$$(k \pm 1)^m = k^m \left(1 \pm \frac{1}{k}\right)^m = k^m \left\{1 \pm \frac{m}{k} + \left[\frac{m(m-1)}{2}\right]\frac{1}{k^2}\right.$$
$$\left. \pm \left[\frac{m(m-1)(m-2)}{6}\right]\frac{1}{k^3} + O\left(\frac{1}{k^4}\right)\right\}, \tag{C.5}$$

then the setting to zero of the coefficients of the terms

$$k^\theta e^{B_0 k}\left(\frac{1}{k^m}\right), \qquad m = (0,1,2,3),$$

240

gives a set of equations that can be solved for (θ, B_0, B_1, B_2). They are

$$B_0 = \pm \cosh^{-1}(A_0) = \ln\left(A_0 \pm \sqrt{A_0^2 - 1}\right), \tag{C.6}$$

$$\theta = \frac{A_1}{\sinh(B_0)}, \tag{C.7}$$

$$B_1 = -\frac{A_2}{\sinh(B_0)} + \left[\frac{\theta(\theta-1)}{2}\right]\tanh(B_0), \tag{C.8}$$

$$B_2 = \frac{[(1-\theta)\cosh(B_0) + \theta(\theta-1)/2 - A_2]B_1}{2\sinh(B_0)} - \frac{A_3}{2\sinh(B_0)} + \frac{\theta(\theta-1)(\theta-2)}{12}. \tag{C.9}$$

The above relations can be rewritten without the use of hyperbolic functions by making the following replacements:

$$\cosh(B_0) = A_0, \tag{C.10}$$

$$\sinh(B_0) = \pm\sqrt{A_0^2 - 1}. \tag{C.11}$$

References

1. J. D. Murray, *Asymptotic Analysis* (Springer-Verlag, New York, 1984).

2. J. Wimp, *Computation with Recurrence Relations* (Pitman, Boston, 1984). See Appendix B.

3. R. E. Mickens and l. Ramadhani, WKB procedures for Schrödinger type difference equations, to appear in *Proceedings of the First World Congress of Nonlinear Analysts* (Tampa, FL; August 19–26, 1992).

BIBLIOGRAPHY

DIFFERENCE EQUATIONS

R. P. Agarwal, *Difference Equations and Inequalities* (Marcel Dekker, New York, 1992).

P. M. Batchelder, *An Introduction to Linear Difference Equations* (Harvard University Press, Cambridge, 1927).

G. Boole, *Calculus of Finite Differences* (Chelsea, New York, 4th edition, 1958).

L. Brand, *Differential and Difference Equations* (Wiley, New York, 1966).

F. Chorlton, *Differential and Difference Equations* (Van Nostrand, London, 1965).

E. J. Cogan and R. Z. Norman, *Handbook of Calculus, Difference and Differential Equations* (Prentice-Hall; Englewood Cliffs, NJ; 1958).

T. Fort, *Finite Differences and Difference Equations in the Real Domain* (Clarendon Press, Oxford, 1948).

C. Jordan, *Calculus of Finite Differences* (Chelsea, New York, 3rd edition, 1965).

W. G. Kelley and A. C. Peterson, *Difference Equations: An Introduction with Applications* (Academic Press, Boston, 1991).

H. Levy and F. Lessman, *Finite Difference Equations* (Macmillan, New York, 1961).

R. E. Mickens, *Difference Equations: Theory and Applications* (Van Nostrand Reinhold, New York, 1990).

K. S. Miller, *An Introduction to the Calculus of Finite Differences and Difference Equations* (Holt, New York, 1960).

K. S. Miller, *Linear Difference Equations* (W. A. Benjamin, New York, 1968).

L. M. Milne-Thomson, *The Calculus of Finite Differences* (Macmillan, London, 1960).

C. H. Richardson, *An Introduction to the Calculus of Finite Differences* (Van Nostrand, New York, 1954).

J. T. Sandefur, *Discrete Dynamical Systems: Theory and Applications* (Clarendon Press, Oxford, 1990).

M. R. Spiegel, *Calculus of Finite Differences and Difference Equations* (McGraw-Hill, New York, 1971).

NUMERICAL ANALYSIS: THEORY AND APPLICATIONS

F. S. Acton, *Numerical Methods that Work* (Mathematical Association fo America; Washington, DC; 1990).

M. B. Allen III, I. Herrerra and G. F. Pinder, *Numerical Modeling in Science and Engineering* (Wiley-Interscience, New York, 1988).

J. B. Botha and G. F. Pinder, *Fundamental Concepts in the Numerical Solutions of Differential Equations* (Wiley-Interscience, New York, 1983).

J. C. Butcher, *The Numerical Analysis of Ordinary Differential Equations: Runge-Kutta and General Linear Methods* (Wiley-Interscience, Chichester, 1987).

E. Gekeler, *Discretization Methods for Stable Initial Value Problems* (Springer-Verlag, Berlin, 1984).

D. Greenspan, *Discrete Models* (Addison-Wesley; Reading, MA; 1973).

D. Greenspan and V. Casulli, *Numerical Analysis for Applied Mathematics, Science, and Engineering* (Addison-Wesley; Redwood City, CA; 1988).

F. B. Hildebrand, *Finite-Difference Equations and Simulations* (Prentice-Hall; Englewood Cliffs, NJ; 1968).

M. Holt, *Numerical Methods in Fluid Dynamics* (Springer-Verlag, Berlin, 1984).

M. K. Jain, *Numerical Solution of Differential Equations* (Wiley, New York, 2nd edition, 1984).

L. Lapidus and G. F. Pinder, *Numerical Solution of Partial Differential Equations in Science and Engineering* (Wiley-Interscience, New York, 1982).

V. Lakshmikantham and D. Trigiante, *Theory of Difference Equations: Numerical Methods and Applications* (Academic Press, Boston, 1988).

W. J. Lick, *Difference Equations from Differential Equations* (Springer-Verlag, Berlin, 1989).

W. E. Milne, *Numerical Solution of Differential Equations* (Dover, New York, 1970).

J. M. Ortega and W. G. Poole, Jr., *An Introduction to Numerical Methods for Differential Equations* (Pitman; Marshfield, MA; 1981).

D. Potter, *Computational Physics* (Wiley-Interscience, Chichester, 1973).

R. D. Richtmyer and K. W. Morton, *Difference Methods for Initial-Value Problems* (Interscience, New York, 1967).

P. J. Roache, *Computational Fluid Dynamics* (Hermosa Publishers; Albuquerque, NM; 1976).

G. D. Smith, *Numerical Solution of Partial Differential Equations: Finite Difference Methods* (Clarendon Press, Oxford, 2nd edition, 1978).

H. J. Stetter, *Analysis of Discretization Methods for Ordinary Differential Equations* (Springer-Verlag, Berlin, 1973).

J. C. Strikwerda, *Finite Difference Schemes and Partial Differential Equations* (Wadsworth and Brooks/Cole; Belmont, CA; 1989).

R. Vichnevetsky and J. B. Bowles, *Fourier Analysis of Numerical Approximations of Hyperbolic Equations* (Society for Industrial and Applied Mathematics, Philadelphia, 1982).

J. Wimp, *Computation with Recurrence Relations* (Pitman, Boston, 1984).

PUBLICATIONS OF MICKENS ON NONSTANDARD SCHEMES

1. "Exact finite difference schemes for the nonlinear unidirectional wave equation." *Journal of Sound and Vibration* **100**, 452 (1985).

2. "Periodic solutions of second-order nonlinear difference equations containing a small parameter III: Perturbation theory." *Journal of the Franklin Institute* **321**, 39 (1986).

3. "Exact solutions to difference equation models of Burgers' equation." *Numerical Methods for Partial Differential Equations* **2**, 213 (1986).

4. "A computational method for the determination of the response of a linear system." *Journal of Sound and Vibration* **112**, 183 (1986).

5. "Mathematical modeling of differential equations by difference equations," in *Proceedings of the First IMAC Conference on Computational Acoustics*, editors, D. Lee, R. L. Steinberg, and M. H. Schultz (North-Holland, Amsterdam, 1987), pp. 387–393.

6. "Runge-Kutta schemes and numerical instabilities: The Logistic equation," in *Differential Equations and Mathematical Physics*, editors, I. Knowles and Y. Saito (Springer-Verlag, Berlin, 1987), pp. 337–341.

7. "Periodic solutions of second-order nonlinear difference equations containing a small parameter IV: Multi-discrete-time method." *Journal of the Franklin Institute* **324**, 263–271 (1987).

8. "An explicit finite difference scheme for linear inhomogeneous hyperbolic equations," in *Contributions in Mathematics and Natural Sciences*, editors, H. W. Jones and C. B. Subrahmanyam (Florida A and M University; Tallahassee, FL; 1986), pp. 147–152.

9. "Pitfalls in the numerical integration of differential equations," in *The Proceedings of the International Workshop on Analytical Techniques and Material Characterization*, editors, W. E. Collins, B.V. R. Chowdari, and S. Radhakrisha (World Scientific Publishing Company, Singapore, 1987), pp. 123–143.

10. "Difference equation models of differential equations." *Mathematics and Computer Modelling* **11**, 528 (1988).

11. "Properties of finite-difference models of non-linear conservative oscillators." *Journal of Sound and Vibration* **124**, 194 (1988).

12. "A difference equation model of the Duffing equation." *Journal of Sound and Vibration* **130**, 509 (1989); with O. Oyedeji and C. R. McIntyre.

13. "Stable explicit schemes for equations of Schrödinger type." *Physical Review* **39A**, 5508 (1989).

14. "Investigations on exact discrete models of continuous systems: Elimination of instabilities," in *Proceedings of the First Edward Bouchet International Conference on Physics and Technology*, editors, L. E. Johnson and J. A. Johnson, III (ICTP; Trieste, Italy; 1988), pp. 219–237.

15. "Finite difference models of ordinary differential equations: Influence of denominator functions." *Journal of the Franklin Institute* **327**, 143 (1990).

16. "Exact solutions to a finite-difference model of a nonlinear reaction-advection equation: Implications for numerical analysis." *Numerical Methods for Partial Differential Equations* **5**, 313 (1989).

17. "Investigation of finite-difference models of the van der Pol equation," in *Differential Equations and Applications*, editor A. R. Aftabizadeh (Ohio University Press; Columbus, OH; 1989), pp. 210–215.

18. "Construction of stable explicit finite-difference schemes for Schrödinger type differential equations," in *Computational Acoustics, Volume I: Ocean-Acoustic Models and Supercomputing*, editors, D. Lee, A. Cakmak, and R. Vichnevetsky (North-Holland, Amsterdam, 1990), pp. 11–16.

19. "A discrete model of a modified Burgers' partial differential equation." *Journal of Sound and Vibration* **142**, 536 (1990); with J. Shoosmith.

20. "Novel explicit finite-difference schemes for time-dependent Schrödinger equations." *Computer Physics Communications* **63**, 203 (1991).

21. "Analysis of a new finite-difference scheme for the linear advection-diffusion equation." *Journal of Sound and Vibration* **146**, 342 (1991).

22. "Construction of a novel finite-difference scheme for a nonlinear diffusion equation." *Numerical Methods for Partial Differential Equations* **7**, 299 (1991).

23. "Nonstandard finite-difference schemes for partial differential equations." *Society for Computer Simulation Transactions* **8**, 109 (1991).

24. "A new finite-difference scheme for Schrödinger type partial differential equations," in *Computational Acoustics*, Volume 2, editors, D. Lee, R. Vichnevetsky, and A. R. Robinson (North-Holland, Amsterdam, 1993), pp. 233–239.

25. "Finite-difference scheme for the numerical solution of the Schrödinger equation." *Physical Review* **A39**, 5508 (1992); with I. Ramadhani.

26. "Finite-difference schemes having the correct linear stability properties for all finite step-sizes," in *Ordinary and Delay Differential Equations*, editors, J. Wiener and J. K. Hale (Longman, London, 1992), pp. 139–143.

246

27. "Finite-difference schemes having the correct linear stability properties for all finite step-sizes II." *Dynamic Systems and Applications* 1, 329 (1992).

28. "Novel finite-difference schemes for partial differential equations," in *IMACS International Symposium on Scientific Computing and Mathematical Modeling*, editors, S. K. Dey and E. J. Kansa (International Journal Services, Inc.; Bangalore, India; December 1992), pp. 136–151.

29. "Finite-difference schemes having the correct linear stability properties for all finite step-sizes III." *Computers and Mathematics* (accepted for publication).

30. "Energy conserving finite-difference schemes for a mixed-parity oscillator." Clark Atlanta University, Center for Theoretical Studies of Physical Systems, Preprint (1993).

31. "A new finite-difference scheme for the Fisher partial differential equation." Clark Atlanta University, Center for Theoretical Studies of Physical Systems, Preprint (1993).

INDEX

www.ingramcontent.com/pod-product-compliance
Lightning Source LLC
Chambersburg PA
CBHW050638190326
41458CB00008B/2319